SpringerBriefs in Applied Sciences and Technology

SpringerBriefs in Computational Intelligence

Series Editor

Janusz Kacprzyk, Systems Research Institute, Polish Academy of Sciences, Warsaw, Poland

SpringerBriefs present concise summaries of cutting-edge research and practical applications across a wide spectrum of fields. Featuring compact volumes of 50 to 125 pages, the series covers a range of content from professional to academic.

Typical publications can be:

- A timely report of state-of-the art methods
- An introduction to or a manual for the application of mathematical or computer techniques
- A bridge between new research results, as published in journal articles
- A snapshot of a hot or emerging topic
- An in-depth case study
- A presentation of core concepts that students must understand in order to make independent contributions

SpringerBriefs are characterized by fast, global electronic dissemination, standard publishing contracts, standardized manuscript preparation and formatting guidelines, and expedited production schedules.

On the one hand, **SpringerBriefs in Applied Sciences and Technology** are devoted to the publication of fundamentals and applications within the different classical engineering disciplines as well as in interdisciplinary fields that recently emerged between these areas. On the other hand, as the boundary separating fundamental research and applied technology is more and more dissolving, this series is particularly open to trans-disciplinary topics between fundamental science and engineering.

Indexed by EI-Compendex, SCOPUS and Springerlink.

SpringerBriefs in Computational Intelligence are a series of slim high-quality publications encompassing the entire spectrum of Computational Intelligence. Featuring compact volumes of 50 to 125 pages (approximately 20,000-45,000 words), Briefs are shorter than a conventional book but longer than a journal article. Thus Briefs serve as timely, concise tools for students, researchers, and professionals.

Samiksha Shukla • Joseph Varghese Kureethara
Jossy P George • Kapil Tiwari

Society 5.0

Journey Towards a Human-Centric World

Springer

Samiksha Shukla
Computer Science and Engineering
Global Academy of Technology
Bengaluru, Karnataka, India

Joseph Varghese Kureethara
Department of Mathematics
Kuriakose Elias College
Mannanam, Karnataka, India

Jossy P George
Computer Science
Christ University
Bangalore, Karnataka, India

Kapil Tiwari
Bangalore, Karnataka, India

ISSN 2191-530X ISSN 2191-5318 (electronic)
SpringerBriefs in Applied Sciences and Technology
ISSN 2625-3704 ISSN 2625-3712 (electronic)
SpringerBriefs in Computational Intelligence
ISBN 978-3-032-22163-6 ISBN 978-3-032-22164-3 (eBook)
https://doi.org/10.1007/978-3-032-22164-3

This Springer imprint is published by the registered company Springer Nature Switzerland AG
The registered company address is: Gewerbestrasse 11, 6330 Cham, Switzerland

Preface

Welcome to *Society 5.0*: *A Journey Towards a Human-Centric World*, a book that invites you to explore a bold vision for the future—one where technology and humanity converge to create a world that is not only smarter but also kinder, more equitable, and deeply sustainable. As we stand at the cusp of a new era, Society 5.0 represents a paradigm shift, redefining how we live, work, and connect in a world shaped by unprecedented technological advancements and a renewed commitment to human well-being.

This book is not just about technology; it is about people. Society 5.0, as envisioned by its pioneers, is a human-centric framework that harnesses the power of artificial intelligence, the Internet of Things, blockchain, and other transformative innovations to address the pressing challenges of our time. From sustainable agriculture to personalized healthcare, from smart cities to inclusive education, this journey is about building a world where no one is left behind—a world where technology serves as a tool for empowerment, not division.

In the chapters that follow, we will trace the evolution of human societies from their earliest forms to the interconnected, data-driven world of Society 5.0. We will delve into the technological infrastructure that forms its backbone, explore its transformative applications across industries, and confront the ethical and societal challenges that come with such profound change. We will also look to the future, envisioning how global collaboration, innovative policies, and emerging technologies can pave the way for a sustainable and equitable tomorrow.

This book is for dreamers, innovators, policymakers, and citizens alike—anyone who believes in the potential of technology to amplify human potential while preserving the planet we call home. As you turn these pages, we invite you to imagine a world where progress is measured not just by efficiency or profit, but by the quality of life it brings to every individual. Let us embark on this journey together, towards a future where humanity and technology walk hand in hand.

Bengaluru, Karnataka, India — Samiksha Shukla
Mannanam, Karnataka, India — Joseph Varghese Kureethara
Bengaluru, Karnataka, India — Jossy P George
Bengaluru, Karnataka, India — Kapil Tiwari

Road Map

Chapter 1: Understanding Society 5.0

This chapter introduces Society 5.0, a vision for a human-centric, tech-driven future. It traces the evolution from Society 1.0 (hunter-gatherer) to 5.0, emphasizing core principles like inclusivity and sustainability. The chapter explores how technologies such as AI and IoT integrate with human needs, prioritizing people over profit. It highlights the holistic interconnectedness of Society 5.0, where digital and physical worlds merge to solve global challenges. Through clear explanations and historical context, readers gain a foundational understanding of how this paradigm shift redefines societal progress, setting the stage for a deeper exploration of its technological and social implications.

Chapter 2: Technological Infrastructure and Framework for Society 5.0

This chapter delves into the technological backbone of Society 5.0, outlining its framework for a connected world. It explores advanced digital infrastructure, including IoT, sensor networks, AI, machine learning, blockchain, and quantum computing. Each technology's role in enabling transparency, security, and efficiency is examined, with practical scenarios illustrating their impact. The chapter emphasizes how these innovations create a robust ecosystem that supports human-centric goals. Through case studies and models, it demonstrates how these technologies integrate to drive smart solutions across industries. Readers will understand the technical foundations that make Society 5.0 possible, balancing accessibility with technical depth.

Chapter 3: Innovations for a Greener Society

This chapter explores Society 5.0's commitment to sustainability, focusing on green energy, smart cities, and eco-friendly systems. It covers renewable energy, sustainable agriculture, water management, and food production, alongside eco-conscious transportation. Governance and policies for sustainable development are analyzed, emphasizing social equity as a cornerstone of a sustainable society. The chapter illustrates how Society 5.0 integrates technology to address environmental challenges while fostering inclusive growth. Through practical examples, it highlights innovative solutions that balance human needs with planetary health, inspiring readers to envision a future where technology and sustainability coexist harmoniously for global well-being.

Chapter 4: Applications of Society 5.0 Framework

This chapter showcases the transformative applications of Society 5.0 across various sectors. It covers Education 5.0, focusing on lifelong learning; Smart Cities, emphasizing sustainable urban planning; Industry 5.0, highlighting human-robot collaboration; Agriculture 5.0, exploring precision farming; and Medical 5.0, delving into personalized healthcare and robotics. Manufacturing 5.0 is examined through smart factories and cobots. Real-world examples illustrate how these applications enhance efficiency, sustainability, and quality of life. The chapter demonstrates how Society 5.0's framework integrates technology to address societal needs, offering readers a vivid picture of its practical impact across global industries and communities.

Chapter 5: Opportunities and Challenges in Society 5.0

This chapter examines the dual nature of Society 5.0, addressing its challenges and opportunities. Challenges include ethical dilemmas, data privacy concerns, the digital divide, policy gaps, and ecological impacts. Opportunities highlight improved quality of life, personalized healthcare, inclusive growth, smart infrastructure, and advanced manufacturing. Through a balanced lens, the chapter explores how these hurdles and prospects shape the adoption of Society 5.0. Real-life examples underscore the stakes, from ensuring equitable access to technology to mitigating environmental risks. Readers gain insight into navigating the complexities of a tech-driven world while maximizing its potential for human benefit.

Competing Interests The authors have no competing interests to declare that are relevant to the content of this manuscript.

About the Book

Society 5.0: A Journey Towards a Human-Centric World presents a fascinating vision of a future where technology works for the benefit of humans. This book presents the evolution of societies in today's interconnected, data-driven world and explains how Society 5.0 integrates AI, IoT, blockchain, and other emerging technologies to create inclusive, sustainable systems. Through different examples and case studies, it explores transformative applications in smart cities, precision agriculture, personalized healthcare, manufacturing, and education, while addressing ethical concerns, data privacy, and the digital divide. Written in an engaging and accessible style, it is a guide for innovators, policymakers, students, academics, and anyone interested in the future of technology. This book inspires readers to imagine and shape a world where innovation enhances human potential and builds a better society.

Contents

About the Authors

Jossy P George is a researcher, author, and academic leader in the field of Computer Science and Human Resources. With a doctoral degree in Computer Science from Christ University, Bengaluru, his research primarily focuses on biometric recognition algorithms, including fingerprint and face recognition, aiming to enhance accuracy and efficiency in biometric systems. He has authored 100 Scopus-indexed publications in reputable international and national journals, contributing significantly to image processing, biometrics, and artificial intelligence. As an accomplished author, Dr. George has published several books, including *Boost Your Ability to Make Great Decisions* and *Leadership Par Excellence*, which aim to motivate and guide readers across various domains. His collaborative work, "Data Ethics and Challenges," addresses contemporary issues in technology and society. Beyond research and writing, Dr. George has been instrumental in advancing higher education, serving as the Director of Christ University's Bannerghatta Campus. His innovative approach, leadership in academic development, and commitment to skill enhancement have earned him recognition as a visionary leader. An active member of international professional associations, he continues to inspire advances in biometric technologies and education reform worldwide.

Joseph Varghese Kureethara is a mathematician and academic leader with expertise in graph theory, optimization, and social networks. Currently serving as Head of the Department of Mathematics at Kuriakose Elias College, Mannanam, India, he has also held positions, such as Treasurer of the Academy of Discrete Mathematics and Applications and Director of the Centre for Research at Christ University. With a PhD in Mathematics from Manonmaniam Sundaranar University, he has made significant contributions through research and publications across data science, education, and network theory. His scholarly works include books on data security, pedagogical innovations, and graph-theoretic concepts, as well as numerous journal articles and conference proceedings.

Dr. Kureethara is an active member of numerous professional societies, including the Indian Mathematical Society, the Ramanujan Mathematical Society, and the Society for Special Functions and Their Applications. His research interests focus on discrete mathematics, social network analysis, and applied optimization, with an emphasis on their roles in education, technology, and societal development. With a career dedicated to advancing mathematical sciences and fostering academic excellence, he continues to inspire students and researchers with his innovative work and leadership in mathematical education and research policy.

Samiksha Shukla (IEEE and ACM Senior Member) is a scholar and educator specializing in data science, cybersecurity, and privacy-preserving technologies. Currently serving as Professor and Dean, Global Academy of Technology, Bengaluru, she has made significant contributions to advancing data science education and research. Dr. Shukla established the Department of Data Science at Christ University's Pune Lavasa Campus, demonstrating her leadership in curriculum development and institutional growth.

Her research focuses on privacy-preserving data mining, secure multi-party computation, and the integration of data science applications in healthcare and social security. She has authored and edited notable publications, including books on data security, data economy, data ethics, and bio-inspired computational paradigms, and has numerous research articles published in reputed international journals and conferences. She is a passionate academician and researcher serving as a reviewer for the Inderscience Journal, Springer Nature's *International Journal of Systems Assurance Engineering and Management* (IJSAEM), and for IEEE and ACM conferences.

Dr. Shukla is actively involved in academic leadership, organizing international conferences in collaboration with Springer, and serving on expert committees. She has been recognized for her contributions to the field with awards, such as the Best Women Faculty Award and university medals. She promotes industry-academia partnerships through collaborations with leading organizations, such as AWS Academy, Google Educator, and Salesforce, fostering practical training and research opportunities for students. As a committed educator and researcher, Dr. Shukla continues to influence the fields of data science and cybersecurity through innovative teaching, impactful research, and active participation in professional organizations, including ACM, IEEE, WiBD, WiDS, and the Computer Society of India.

Kapil Tiwari is a Senior Manager of Software Development at Amazon with over 20 years of extensive experience in AI, big data, and privacy-preserving machine learning. Holding a PhD in Computer Science, he specializes in developing innovative solutions that integrate ethical considerations with cutting-edge technology. His research contributions include multiple patents, influential publications, and books on data ethics and the digital economy, positioning him as a thought leader in the field. Throughout his career, Dr. Tiwari has led diverse teams in building scalable, AI-powered automation and analytics platforms using advanced tools, such as AWS, Amazon Bedrock, and Data Mesh. His expertise spans web services, Angular, and modern cloud architectures, with significant achievements in developing enterprise-grade Office Add-ins and document management solutions. Recognized for his innovative approach, he has received awards, including the EMC Rock Star Award and the EMC Excellence Gold Award.

Beyond industry, Dr. Tiwari actively contributes to academia as a member of university boards of studies, shaping the next generation of computer scientists. An avid keynote speaker, he frequently participates in industry and academic conferences, advocating for responsible AI and data ethics. His work embodies a commitment to driving technological progress while upholding ethical standards, shaping the future of AI and data science at the intersection of innovation and ethics.

Chapter 1
Understanding Society 5.0

1.1 Society 5.0

Society 5.0 is an innovative concept developed by the Japanese government that envisions a super-smart society [1] in which advanced technologies, especially AI and IoT, are seamlessly integrated into every facet of life. This paradigm shift aims to build a sustainable and inclusive society that balances economic development with the addressing of critical social issues. Society 5.0 aims to create smart cities, enhance wellness, streamline energy use, and improve quality of life by utilizing big data and cutting-edge technologies, thereby overcoming the limitations of previous industrial revolutions, such as limited personalization and the lack of environmental and social sustainability, as central goals in Industry 4.0. The current approach centers on human-centric development, personalization, and sustainable environmental practices to foster innovation that benefits all sectors of society and reduces disparities. The ultimate goal is to ensure that technological progress improves well-being and environmental sustainability, promoting overall societal harmony. Society 5.0 represents a holistic view of the future, in which technology serves human needs and collaborates in positive transformation. This chapter presents a framework that harnesses cutting-edge technologies, including artificial intelligence, digital twins, data analytics, and robotics, to support collaborative creation, customization, personalized experiences, and co-production.

1.2 The Evolution of Societies: From 1.0 to 5.0

On January 22, 2016, the Government of Japan revealed its pioneering vision for the future with the release of the Fifth Science and Technology Basic Plan. This plan presents the concept of "Society 5.0", a "futuristic model of a society guided by

S. Shukla et al., *Society 5.0*, SpringerBriefs in Applied Sciences and Technology, https://doi.org/10.1007/978-3-032-22164-3_1

scientific and technological innovation". The primary objective of Society 5.0 is to integrate the physical and digital worlds by harnessing the full potential of Information and Communication Technologies (ICTs). The ultimate goal is to create a "super-smart society" that will bring wealth and well-being to the people. This innovative plan represents a significant shift in Japan's approach to shaping its future. Rather than focusing exclusively on economic growth, the government has recognized the transformative power of emerging technologies and their ability to address societal challenges. By combining the digital and physical worlds, Society 5.0 aims to create an efficient, sustainable, and inclusive development model. The plan's emphasis on scientific and technological innovation presents Japan's commitment to positioning itself at the forefront of the global knowledge economy. As the world looks to the future, Society 5.0 serves as a confident roadmap for navigating the complex landscape of the twenty-first century.

As illustrated in Fig. 1.1, the super-smart society has undergone several revolutionary phases. The transformations navigated by scientific and technological innovation have given rise to a new form of society, surpassing previous stages of human civilization. Following the eras of hunter-gatherer, agricultural, industrial, and information societies, we are entering the emergence of "Society 5.0," a vision championed by the government of Japan.

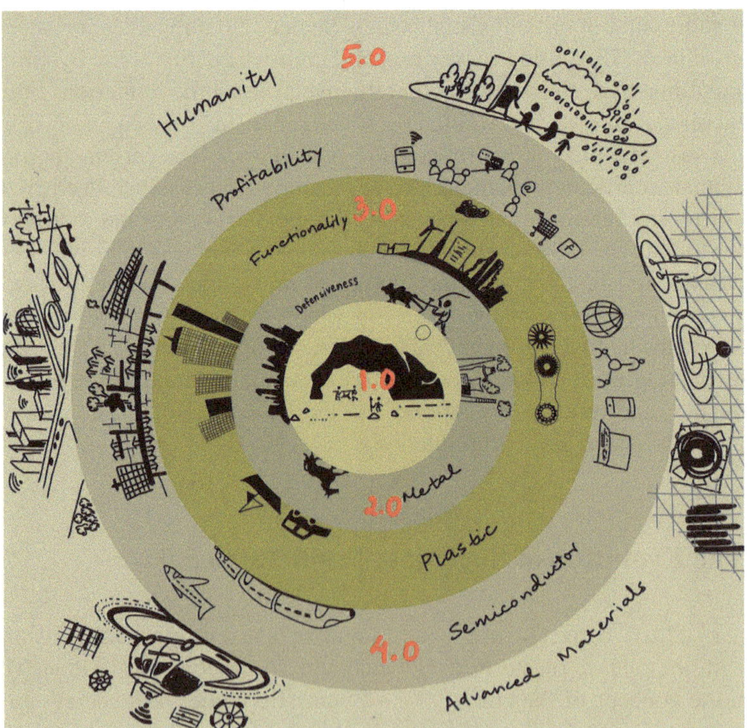

Fig. 1.1 Revolution of society

In Society 5.0, the cyber-physical worlds converge seamlessly, which is enabled by the full potential of Information and Communication Technologies (ICTs). This new societal model aims to deliver exceptional wealth and well-being to the people by integrating innovation with a focus on addressing societal challenges. As it navigates the challenges of the twenty-first century, Society 5.0 offers a bold and ambitious pathway forward. One way to consider "The Evolution of Societies: From 1.0 to 5.0" is as a framework for understanding how social structures, technology, governance, and human awareness have evolved across human civilizations. This framework, although absent from the scholarly literature, helps explain how civilizations have changed and evolved.

Society 1.0 consists of tribal and agricultural societies. These societies were active around 3000 BCE. It was formerly a small tribal organization operating within close-knit tribes or clans. Fighter rank, family bonds, or religious position determined leadership. In these societies, individuals should be able to hunt, gather, and farm. Independent societies concealed most communities, limiting business and contact. Religion held paramount importance, interweaving itself with the political structure. Elders and leaders were always viewed as intermediaries between the people and the divine. The most common tools were made of natural materials, primarily stone and wood.

Society 2.0 constitutes the Industrial Revolution, spanning the eighteenth and early twentieth centuries. The agrarian economies gave way to factory production, mechanization, and bulk manufacturing. The defining factors of this society were urbanization and large-scale production. This society emerged as a working-class one, shifting the paradigm from rural and agricultural backgrounds to factory-based, urbanized work. It led to class differentiation. The growth in sectors such as railroads, steam engines, and electricity empowered the economies and industries. As a result, marketplaces have expanded, and global trade has begun to take off.

Society 3.0 is the next stage of human development, in which technology and society interact to improve how individuals live, work, and communicate. With the advent of AI, automation, and the internet, life has become more connected and efficient. Society 3.0 signifies a paradigm shift from conventional centralized power structures toward a cooperative, networked, and sharing-based society. In this approach, individuals prioritize teamwork over competition and sharing over ownership, collaborating to generate value through social innovations and new digital platforms. Technological breakthroughs empower people and communities, enabling informal, decentralized economies to emerge and operate alongside, or independently of, formal economies. Society 3.0 features the rise of "value networks," in which individuals freely exchange opportunities, resources, and expertise and collaborate as flexible teams rather than within strict job structures. Learning is continuous, contextual, and socially created, and it frequently occurs outside traditional institutions. Society 3.0 emphasizes access over possession. This innovative strategy promotes a more inclusive, sustainable, and rewarding way of living.

The next step after Society 3.0 is Society 4.0, in which technology plays a greater role in shaping human lives. The primary focus of Society 4.0 is to use advanced technologies such as IoT, robotics, and AI to make life sustainable, effective, and individualized. For example, self-driving cars use sensors and AI to navigate traffic, making commutes more efficient and safe. It prioritizes innovation, collaboration, and the alignment of technology with human needs. In Society 4.0, almost everything around us is connected and can communicate with each other. For example, in a smart factory, robots work alongside humans, performing tasks such as assembly and quality checks, while AI manages the entire production process to ensure smooth operation.

Japan introduced the Society 5.0 concept, envisioning a human-centered, technology-driven society that merges cyber and physical spaces to improve quality of life. It was built on the foundations of prior societies, with an emphasis on addressing social challenges [2] through data science, AI, robotics, and smart technologies. For example, a smart healthcare system for personalized treatment of older people employs wearable devices for continuous monitoring, AI-driven diagnostics, and robotic caregivers. It facilitates access to state-of-the-art medical facilities in remote locations while reducing the burden on human doctors. This society emphasizes innovation, collaboration, and a balance between technology and human needs, aiming to address significant challenges, including climate change, poverty, and health care, by integrating technology into every aspect of life.

1.3 The Core Principles of Society 5.0

Society 5.0 is an innovative concept that prioritizes people-centric technological advancements. It indicates the future of societal evolution: the integration of the digital and real worlds to address societal issues and improve quality of life. Society 4.0 focuses primarily on data use and digital transformation; Society 5.0 aims to build a highly intelligent society in which technology assists humanity in a sustainable, balanced way. In Fig. 1.2, the fundamental principles reinforcing Society 5.0 ensure that technology development aligns with human needs, ethical values, and sustainability. Despite its features, the AI and automation integral to Society 5.0 raise significant concerns about job displacement, leaving workers vulnerable to unemployment and insecurity. This structural shift demands extensive workforce retraining initiatives to develop specialized digital skills. Fundamentally, Society 5.0 is envisioned as human-centric, aiming to enhance human capabilities and prioritize human welfare over efficiency, thereby reducing the risk of mass exclusion.

1.3.1 Human-Centric Perspective

At the core of Society 5.0 is the principle of prioritizing human well-being in all technological developments. The approach shifts the focus from technology driving the pace of societal change to genuine improvements in individuals' lives enabled by IoT [3], big data, AI, and robotics. The goal is to create an environment that is more inclusive and accessible to everyone. The emphasis is on creating a society in which technological advancements improve the quality of life, ensuring that

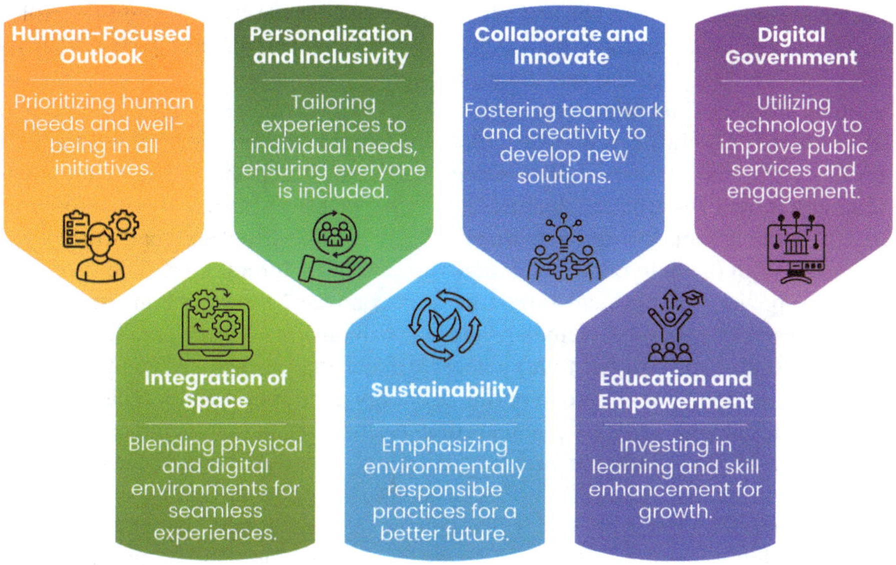

Fig. 1.2 Principles of Society 5.0

everyone can benefit from these innovations. By focusing on ethical values and sustainability, Society 5.0 aims to foster a harmonious coexistence between humans and technology, for instance, an AI-powered wearable medical device. This wearable healthcare device can monitor vital signs and provide real-time health insights, enabling early detection of potential health issues. It demonstrates how Society 5.0 uses technology to enhance human life, maintain ethical standards, and promote sustainability. A human-focused outlook places people, ethics, and social well-being at the center of decision-making, ensuring fair, inclusive, and responsible outcomes. It addresses social bias by actively identifying inequalities and incorporating diverse perspectives to reduce discrimination. Understanding human behavior, emotions, and cultural contexts helps anticipate how individuals and communities may respond to policies, systems, or technologies. This outlook also acknowledges the role of social influence, including norms, peer behavior, and the media, in shaping opinions and actions. Most importantly, it supports informed decision-making by balancing data and evidence with human judgment, empathy, and values, resulting in transparent, trustworthy, and socially sustainable decisions.

1.3.2 Integration of Cyber and Physical Space

In the context of Society 5.0, cyberspace doesn't refer to a space for voluminous data exchange but rather to a space created by computer networks for analyzing problems and tailoring real-world solutions. Society 5.0 proposes a seamless connection between the cyber and physical spaces, driven by advanced technologies. By using advanced tools such as IoT, AI, and big data, it aims to rebuild industries and services, enabling real-time data collection, analysis, and application to optimize processes and improve daily life. For example, smart cities use IoT-enabled devices to manage traffic, prevent deadlocks, minimize congestion, and improve the efficiency of public transport. This integration makes metropolitan life more efficient, easier, and sustainable. A notable example of Society 5.0 is Toyota's Woven City, a "living laboratory" in Japan that skillfully blends virtual and real-world environments. With the help of a hidden underground layer for automated delivery and hydrogen power, the physical infrastructure "connects" three separate street patterns for pedestrians, small vehicles, and self-driving cars. Every physical asset is replicated online as a digital twin, where artificial intelligence (AI) analyzes data from home sensors in real time to optimize energy distribution and forecast maintenance requirements. This deep integration focuses on human-centric outcomes, such as non-invasive in-home health monitoring, where AI detects early signs of illness. Additionally, driverless e-Palette shuttles support an inclusive future by serving as mobile clinics or offices that deliver services directly to senior citizens, thereby bridging the physical-digital divide. By treating an entire community as a programmable interface, Woven City demonstrates that Society 5.0 necessitates a radical redesign of physical space to facilitate the high-speed data flows essential for human well-being.

1.3.3 Personalization and Inclusivity

Society 5.0 offers solutions that address individual needs while promoting inclusivity across all segments, irrespective of age, ability, or location. As a result, everyone can benefit from technological advancements. For example, imagine fitness software that adjusts its training rules based on the user's age, fitness level, and physical restrictions. In addition to offering accessible alternatives for older people or people with disabilities, it provides individualized recommendations. By doing so, it promotes involvement from a wide variety of people, creating an inclusive health and wellness community. Estonia is recognized as the world's first digital republic, offering a pragmatic and practical model of Society 5.0 through its sophisticated e-Estonia ecosystem. The core principle of this system is data sovereignty, which ensures that citizens retain ownership and control over their personal data rather than giving it to the government. Every resident is assigned a secure digital identification (e-ID), which enables smooth access to critical services such as healthcare, banking, taxation, and education. These services are linked via X-Road, a secure data-sharing platform that adheres to the "once-only" principle. Ensuring that citizens provide information only once, while it is securely used across government systems. Estonia implemented i-Voting to promote inclusiveness, enabling citizens to vote remotely and benefiting older people, individuals with impairments, and those living abroad or in rural areas. Transparency is maintained using systems such as the Data Tracker and the KSI Blockchain, which allow citizens to monitor data access. Together, these developments demonstrate how personalized digital governance may coexist with privacy and trust in Society 5.0.

1.3.4 Sustainability

Society 5.0 prioritizes addressing urgent global issues, including resource shortages, global warming, and the challenges presented by aging populations, while fostering economic growth. This method aims to establish a harmonious connection between environmental conservation and technological development. For example, a corporation may develop smart agricultural technology to address resource shortages, thereby optimizing water use and reducing water waste. By employing sensors and data analytics, farmers can monitor soil conditions and crop health in real time, leading to more effective farming practices. It boosts crop yields, benefiting local economies and promoting resource conservation. The government can make these technologies available to elderly farmers, enabling everyone to benefit from sustainable agricultural techniques. The Spread Smart Farm in Kameoka, Japan, is a prime real-world example of Society 5.0, combining cyber and physical technologies to promote sustainable, human-centric agriculture. The facility functions as a powerful Cyber-Physical System (CPS), with thousands of IoT sensors monitoring plant health, CO_2 levels, and nutrient concentrations. This data is analyzed in cyberspace by AI algorithms that compute optimal light and nutrition "recipes" while forecasting harvest dates with 99% precision. Automated robots then implement these digital instructions by precisely sowing, watering, and harvesting. This "Techno Farm" achieves high ecological sustainability by recycling 98% of the water used and eliminating chemical pesticides through a strictly controlled "clean room" environment. Vertical gardening yields 30,000 lettuce heads each day on just a small portion of the land required for conventional soil-based cultivation. Crucially, the farm exemplified Agriculture's 5.0 emphasis on hyper-personalization; for example, it can adjust nutrient levels to produce low-potassium lettuce for people with renal disease. Ultimately, the Spread Smart Farm transforms agriculture from a labor-intensive enterprise into an inclusive, data-driven service that prioritizes human health and environmental sustainability.

1.3.5 Collaborate and Innovate

Society 5.0 encourages collaboration among communities, corporations, and governments to develop creative solutions to a range of societal problems. This collaborative idea harnesses the strengths and resources across different sectors to drive meaningful change. For example, to address homelessness, a city administration may collaborate with nearby IT firms and civic associations. They may jointly develop a mobile application that connects people in need with accessible services such as food banks, shelters, and skill-training courses. These partners may establish a comprehensive support system that helps people achieve stability and independence in addition to meeting their urgent needs by combining their resources and skills. This partnership demonstrates how teamwork can lead to significant solutions to critical societal problems. BMW and Microsoft established the Open Manufacturing Platform (OMP), which represents Society 5.0's Collaborative Innovation by eliminating industrial silos with co-opetition. Companies compete on vehicle design while sharing non-competitive data, such as robotic arm optimizations using open-source frameworks. This architecture serves as a collective Cyber-Physical System (CPS), with anonymized data from thousands of devices pooled into a shared "data lake." Global AI models then examine this different dataset to uncover performance patterns; for example, lessons from BMW's manufacturing processes can be applied to Bosch's production using standardized techniques. This Digital Commons democratizes access to advanced AI by enabling Small and Medium Enterprises (SMEs) to use such technologies that would otherwise be prohibitively costly. Finally, the sources indicate that the OMP promotes sustainability, demonstrating that innovation in a "Super Smart Society" is a collective effort for shared economic and ecological progress.

1.3.6 Education and Human Resource Development

Education is a key factor in the development of a society and culture of mutual respect. It prioritizes equipping individuals with the necessary problem-solving skills and creativity through innovation and exploratory learning methods. In a rapidly evolving digital landscape, education and HR developments play a key role in empowering the workforce to address challenges created by technological advancements. It ensures the workforce is adaptable to change. Promoting a culture of lifelong learning ensures people stay up to date with the skills needed to embrace innovative ideas in today's job market.

Additionally, the emphasis on education fosters innovation by cultivating creative talent, encouraging individuals to think outside the box, and enabling them to contribute to societal advancement. For example, VR-based training programs provide immersive simulations of real-world situations across industries such as construction, manufacturing, and healthcare, enabling skill refinement in a safe environment before applying them in real-world settings. This approach not only enhances learning but also builds confidence, thereby fostering a skilled and adaptable workforce ready to meet the requirements of Society 5.0.

1.3.7 *Digital Government*

Digital government serves as a foundation for Society 5.0, utilizing innovative technologies such as blockchain and AI to promote ethical and transparent governance and citizen-centered policymaking. It significantly reduces administrative delays and facilitates easy access to pivotal public services. It enhances transparency between citizens and the government, thereby building trust among all stakeholders by enabling citizens to understand how the government makes decisions and allocates resources.

Additionally, data-driven policymaking enables leaders to make informed decisions that enhance community well-being and address urgent issues, such as social inequality and environmental sustainability. A notable example is the use of blockchain technology for voting to strengthen election security and transparency, thereby reducing the risk of fraud. By using this tool, digital government not only reforms and digitizes the traditional administrative system but also engages citizens, who feel heard and valued. Ultimately, this approach aims to foster a fair society in which technology enhances human life, leading to an equitable, efficient, and responsive future.

1.4 Technology and Society 5.0

The goal of Society 5.0 is to deploy cutting-edge technologies to address societal issues and improve citizens' quality of life, building a "Super-Smart Society". As mentioned in Sect. 1.1, the government of Japan proposed that Society 5.0 will go beyond the information society by integrating the cyber and physical spaces while prioritizing human needs and values.

In contemporary society, technology is integral to daily life. It goes beyond the offices and industry machinery. Instead, it deeply embeds itself in daily life, changing how we work, travel, live, sustain our health, and interact with government systems. The main objective of Society 5.0 is to build a society in which technology actively supports individuals' comfort, security, and happiness. Figure 1.3 shows how the integration works:

1.4.1 *Data Collection Using IoT Sensors*

Imagine your home, office, city streets, and the neighborhood clinic becoming aware of their surroundings. It is possible with the help of IoT. IoT sensors are small, often barely visible devices embedded in things that continuously collect data. Advanced thermostats in smart home systems monitor temperature and learn user preferences. These sensors can also help factories monitor device temperature

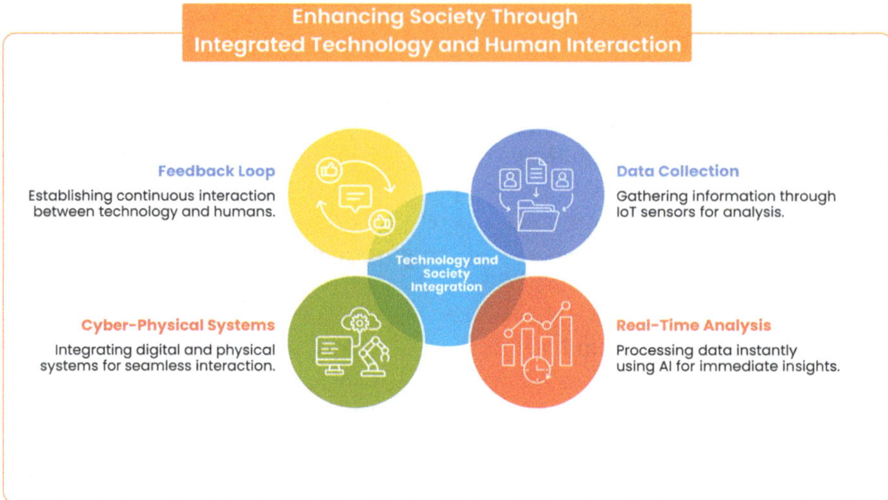

Fig. 1.3 Technology and human synergy

to prevent malfunctions. Wearable devices continuously monitor patients' vital signs and alert medical practitioners when abnormal patterns are detected. In smart cities, these sensors monitor energy use, traffic, and air quality.

The foundation of Society 5.0 is continuous change, which functions as a nervous system that enables technology to sense what's happening. For instance, BMW's Regensburg Open Manufacturing Platform (OMP) operates as an integrated cyber-physical system in which IoT sensors gather real-time data from manufacturing machinery. The data is anonymized and aggregated into shared platforms, where AI algorithms analyze patterns related to efficiency, energy usage, and machine condition. The knowledge and insights produced are subsequently implemented in production operations via automated modifications and predictive maintenance. In addition to enhancing efficiency, OMP facilitates sustainability objectives and provides Small and Medium Enterprises (SMEs) with access to sophisticated digital technologies. OMP demonstrates how collaborative data sharing can drive inclusive, efficient, and sustainable industrial development in Society 5.0.

1.4.2 Real-Time Data Analysis Using AI

Once the data is collected, it is not only stored but also analyzed using AI tools. It turns data from IoT devices into insightful information. Let us examine a few examples in which AI analyzes real-time data. The healthcare industry uses wearable devices to identify early warning signs of illness and alert patients, physicians, and caregivers. In cities, AI is used to manage and analyze traffic patterns, optimizing signal timings and reducing congestion. In agriculture [4], it assesses soil conditions and weather forecasts to recommend optimal planting and harvesting times. In business, analyzing consumer behavior helps strengthen supply chain strategies by addressing issues before they become serious. With real-time data analysis, society can move from a reactive to a proactive mode. An actual instance of this is the Automated Quality Image Recognition system employed by the BMW Group. High-definition cameras on the assembly line serve as sensors. At the same time, AI models evaluate images in real time to identify anomalies in standard components, such as misaligned bolts or surface imperfections, with greater speed and precision than a human inspector. This anonymized analytical data is subsequently uploaded to the OMP "data lake," enabling other users to enhance their computer vision models utilizing these shared patterns.

1.4.3 Cyber-Physical Systems (CPSs)

They integrate processes, real-time data, and computational algorithms through sensors, actuators, and network communications, enabling dynamic decision-making without human intervention. For example, in industrial robotics, arms automatically adjust production processes based on sensor data to maintain precision and minimize faults. At the same time, self-driving vehicles use LiDAR cameras and AI to navigate safely by utilizing real-time traffic data. CPS optimizes energy consumption in smart buildings via a continuous feedback loop. IoT sensors gather real-time data on occupancy, temperature, and HVAC status. AI analyzes the input, predicts energy demand, and determines the minimum power required to maintain human comfort. The system then automatically adjusts physical assets, such as fan speeds, lighting levels, and window shades, with precision. This proactive approach prevents energy waste, resulting in significant cost savings and decreased environmental impact. The Toyota Woven City is a prime example of the integration of cyber-physical systems within Society 5.0.

1.4.4 Feedback Loop

The data feedback loops between humans and technology ensure that humans will remain central to the digital transformation process. Humans interact with these advanced systems through devices such as mobile applications for food delivery, bill payments, healthcare, and transport. Every time people use these applications intentionally or unintentionally, they contribute to the continuous data stream that feeds back into the system. This constant exchange of information allows technology to learn from human behavior, adapt its responses, and provide personalized, efficient, and accurate feedback. This integration provides society with a real-time ecosystem that seamlessly blends human needs and technology, delivering innovative, human-centered solutions and improving quality of life.

By integrating advanced technologies such as AI, IoT, and CPS [5] into daily life, Society 5.0 is transforming key sectors (healthcare, smart homes, transport, etc.) to create a proficient, equitable, and human-centric society. In healthcare, doctors can monitor patients remotely using real-time data and adjust treatment plans. This intervention can prevent hospitalization and acute harm to patients by providing personalized, timely care. Urban planning benefits from this integration, which analyzes traffic, energy use, and population data, thereby optimizing route planning and infrastructure and enhancing the standard of living through a smarter, more sustainable environment. Just-in-time manufacturing and delivery, enabled by AI and IoT, have transformed the supply chain industry by reducing waste and allowing businesses to adapt to changing demand. Digital government platforms provide equitable, tailored services that make public services more inclusive, ensuring no one is left behind. The aim of Society 5.0 is not to replace humans with machines but to enhance human capabilities, comfort, and well-being. Integrating technology into everyday life creates a smarter, more connected ecosystem that addresses social challenges, reveals human potential, and builds a more responsive, fair, and comfortable society.

1.5 Human Centered Design: A People-First Approach

Human Centered Design (HCD) has evolved within the Industry 5.0 framework [2, 6], emphasizing a paradigm shift from traditional user-centered models toward a broader consideration of social, ethical, and environmental dimensions. The main aim of HCD is usability, not only ease of use but also sustainability and resilience. This method draws on psychology, sociology, and anthropology to address complex societal issues and develop tools and technologies that actively promote long-term human welfare in a rapidly changing world. HCD set the ground for sustainable development by fostering strong collaboration between humans and smart technologies, creating spaces where ethical values, creativity, and human intuition blend seamlessly with efficient machines and data analytics. Figure 1.4 demonstrates the human-centered design.

This section presents the evolution of HCD in Industry 5.0, underscoring the transition from conventional user-centric approaches to a broader integration of social, ethical, and environmental preferences.

Here, the focus is on the crucial role of designers, production teams, and end users working together in an interconnected, data-driven design ecosystem that enables real-time feedback, continuous collaboration, and ethical engineering practices. The authentic HCD in Industry 5.0 depends on integrating advanced technologies with a strong commitment to human well-being and social values, thereby enabling more sustainable and ethically responsible practices that reflect the vision of Society 5.0.

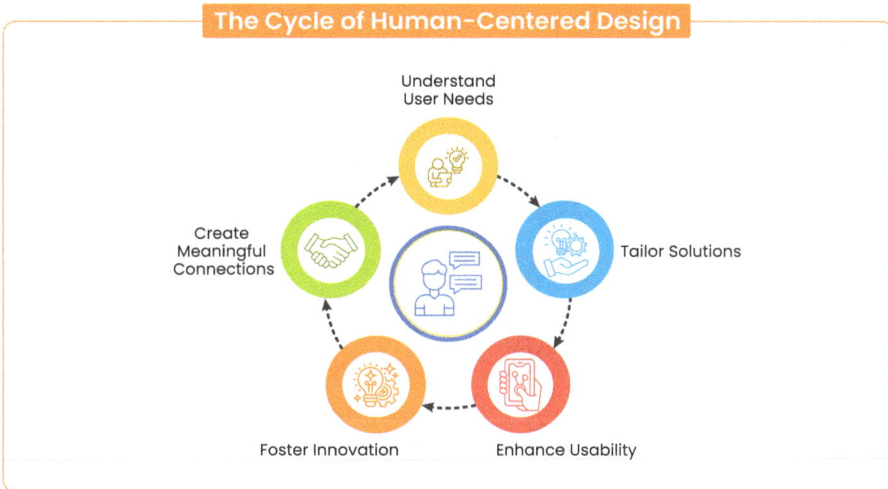

Fig. 1.4 Human centered design

1.5.1 Key Principles of Human-Centered Design (HCD)

It is a creative approach to problem-solving. HCD follows a process that places people at the center at every stage of development. Instead of starting with technology, designers begin by understanding users: their daily activities, real needs, and reasons behind their actions. By closely observing and interacting with people over time, designers gain a fundamental understanding of how people live and work. This understanding helps them create products, systems, and services that fit real life. As a result, the outcomes are helpful, easy to use, and satisfying. In Society 5.0, HCD ensures that designers develop smart technologies to improve human life and address social challenges, rather than forcing people to adapt to technology. The focus remains on meeting human needs first, so progress feels natural and benefits everyone.

It ensures that solutions are designed to meet users' needs, inputs, and actions. This section explains these principles, as shown in Fig. 1.5. The modern approach to building systems and technologies is influenced by earlier methods, mainly agile software development. Agile introduced the idea of working in small, ongoing cycles that develop solutions, test, and improve continuously, rather than following a rigid plan. It allowed teams to respond quickly to change and improve quality over time. Today's design approaches extend this idea by placing greater emphasis on the end user. The focus is no longer limited to producing functional software or systems. Instead, it involves understanding how people live, work, and interact with technology in their daily lives. By considering emotions, comfort, and ease of use, designers aim to create solutions that offer absolute satisfaction, support human well-being, and deliver lasting value to society.

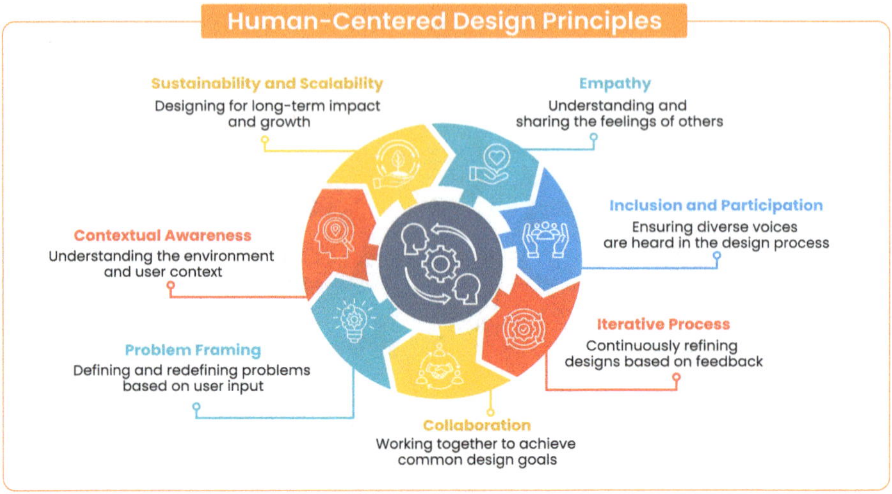

Fig. 1.5 Principles of human centered design

Empathy: At the core of designing with people in mind. Designers pay close attention, observe how people act, and ask thoughtful questions to understand what matters to them. Rather than guessing what users might want, empathetic designers take the time to understand and empathize with the challenges and expectations. This human-centric design approach reveals hidden insights, leading to solutions that are meaningful and enjoyable for users. Empathy guides an appreciative, personal, and reflective experience.

Inclusion and Participation: It involves users at every stage of the design process, from ideation to implementation. HCD prioritizes diverse, often personalized perspectives, particularly from those affected by the solutions. By involving users in brainstorming, prototyping, and testing, designers create outcomes that align with real-world needs and experiences. This collaborative approach unveils hidden challenges, reduces bias, and fosters trust. It also empowers users, giving them a sense of ownership that enhances the acceptance and effectiveness of solutions. When inclusion goes beyond equity, it drives superior design, and when diverse voices shape the process, the results are more innovative, practical, and successful in addressing real-world problems.

Iterative Process: Designers advance iteratively by creating, testing, and continuously improving. Instead of waiting to perfect an idea before sharing it, designers build simple prototypes and test them with users. These early models might be sketches, prototypes, or basic working versions. User feedback helps identify what works and what does not. Designers then refine their ideas and repeat the cycle. This process reduces risk, saves time, and ensures the final solution is effective. Iteration transforms good ideas into great ones by learning from mistakes and insights along the way. It's a loop of learning, not a straight path.

Collaboration: HCD promotes teamwork across diverse disciplines, connecting varied expertise to spark innovation. By weaving together diverse perspectives, HCD encourages the development of innovative solutions that address a broad range of user needs. This collaborative approach breaks down silos, enabling teams to blend insights from interdisciplinary fields and achieve more inclusive and effective outcomes. When individuals from diverse backgrounds collaborate, they bring unique perspectives that enrich the design process, ensuring that solutions are not only creative but also resonate deeply with users. By prioritizing cooperation, HCD cultivates an environment where diverse ideas flourish, leading to products and services that serve a broad audience.

Problem Framing and Reframing Based on User Input: HCD flourishes on interdisciplinary collaboration and continuous user engagement. By bringing diverse teams together, HCD delivers creative solutions that address a broad spectrum of user requirements, blending unique perspectives to promote inclusive innovation. Continuous user feedback is essential for guiding designers to understand what works, what falls short, and where improvements are required, thereby ensuring that the final product aligns with user expectations. Equally critical is problem framing and reframing. By listening to user inputs and exploring their context, designers uncover deeper challenges and unexpected opportunities. This approach drives innovation, ensuring efforts target what truly matters.

Contextual Awareness: In crafting solutions within Human-Centered Design, understanding the user's world is essential. Design flourishes in daily life experience, shaped by cultural beliefs, economic conditions, social expectations, infrastructure, and access to technology. For example, an application designed for rural communities with limited internet access must function offline and be easy to use. By prioritizing context, designers ensure that solutions are practical, respectful, and meaningful while avoiding unintended pitfalls. Immersing oneself in the user's environment and understanding its nuances and constraints enables the creation of products that seamlessly integrate into lives, meeting fundamental needs with precision and care and fostering genuine connection and impact in the spaces where people live and thrive.

Sustainability and Scalability: Crafting solutions that ensure sustainability and scalability ensures continuous impact and growth. A one-time, single-context fix is insufficient; solutions must evolve and reach broader audiences. It demands early consideration of resources, maintenance, user education, and adaptability. Sustainable design honors the environment, communities, and long-term effects, while scalable solutions are flexible enough to adapt to new settings or larger groups. By blending these principles, designers create products that go beyond their debut and deliver lasting value. Thoughtful planning ensures that solutions remain relevant, resilient, and impactful, transforming lives and environments with meaningful, far-reaching benefits in an ever-evolving world.

1.5.2 *Human-Centered Design Process*

The human-centered design process involves the following stages, as shown in Fig. 1.6.

Research: This stage involves a deep understanding of the target audience to signal the design process. Designers conduct interviews, surveys, and field studies to gather qualitative and quantitative data about users' behaviors, needs, and challenges. It may include observing users in their natural environments or analyzing existing data to uncover patterns. The goal is to empathize with users, identifying their motivations and concerns without making assumptions. By combining insights, designers build a clear picture of the audience's context, ensuring that the design addresses real-world problems. Effective research lays the foundation for all subsequent stages, fostering innovation customized to specific needs.

Define: In this stage, designers analyze research findings to pinpoint the core problems and user requirements. It involves examining data from interviews, surveys, and observations to identify patterns and key insights. Designers develop tools such as personas, empathy maps, and problem statements to articulate the user's perspective and needs clearly. The goal is to frame the challenges in a way that guides the design process and ensures user-focused solutions. By narrowing the scope, this stage transforms raw data into actionable insights, providing a clear direction for ideation. A well-defined problem statement aligns the team and ensures that the design effectively addresses the most critical user requirements.

Ideate: This stage triggers creativity to explore diverse possibilities and to develop personalized solutions aligned with the defined requirements. Assemble the team to brainstorm freely, encouraging bold and unconventional ideas without judg-

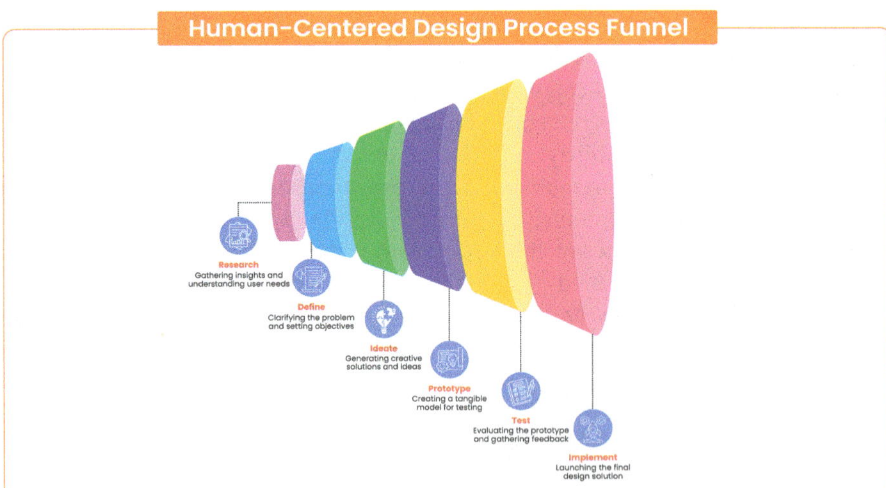

Fig. 1.6 Human-centered design process

ment. Capture every thought, from practical to unconventional, to ensure a broad spectrum of options. Use tools, such as mind maps or collaborative workshops, to generate inspiration and identify innovative approaches. The goal is to create a diverse collection of ideas that effectively addresses the challenge, setting the foundation for further development. This stage fosters open dialogue and inclusivity, ensuring that all perspectives are considered when designing solutions that resonate deeply with the target audience.

Prototype: In this stage, the ideas take tangible form through simple prototypes, transforming abstract concepts into visual and interactive representations. Create simple sketches, wireframes, or basic models to bring ideas to life, keeping the process agile and cost-effective. These prototypes act as primary drafts, allowing teams to test assumptions and explore functionality without a huge investment. Whether using paper models or digital tools, the focus is on clarity and experimentation. Prototypes help teams visualize potential solutions, identify flaws, and refine concepts early on. This stage encourages quick iteration, enabling designers to adapt and evolve ideas based on initial insights, paving the way for user-centered refinements.

Test: This stage involves engaging users to interact with prototypes and provide valuable feedback. By observing how real users navigate and respond to the proposed designs, teams can present strengths, weaknesses, and unpredictable insights. Conduct usability tests, interviews, or surveys to gather honest input, with a focus on user experience and functionality. This stage is critical for validating ideas and identifying areas for improvement. Iterating based on feedback ensures the solution aligns with user needs and expectations.

Implement: In the final stage, designers refine the design using insights gathered from testing to create a polished, market-ready product. Incorporate feedback to address flaws, enhance usability, and ensure alignment with user needs. Finalize all elements, from aesthetics to functionality, while preparing for a smooth launch. This stage involves collaboration across teams to resolve outstanding issues, conduct quality checks, and establish a launch plan. Documentation, user guides, and training need to be developed to support adoption. The goal is to deliver a solution that not only meets expectations but also enhances user experience, ensuring a successful transition from concept to reality with an enduring impact.

1.5.3 Benefits of HCD

HCD transforms product development by placing users at the core. This approach crafts solutions that align closely with user needs, fostering greater satisfaction and delivering experiences that feel simple and rewarding. Products designed with HCD are readily adopted because they connect deeply with users, facilitating seamless integration into their lives. By engaging users early through iterative testing, HCD uncovers potential flaws before they become issues. They escalate problems early in the design process, saving significant time and resources later.

Furthermore, its collaborative process sparks creativity, yielding innovative designs that distinguish themselves in competitive markets. Ultimately, HCD ensures creations are not only functional but also emotionally resonant, building meaningful connections with users. By prioritizing empathy and iteration, this design philosophy drives impactful, user-focused solutions that elevate both utility and engagement, redefining success in design.

1.6 The Interconnectedness of Society 5.0

Society 5.0 envisions a future in which digital innovation and human well-being blend to create a balanced, inclusive world. Unlike past industrial revolutions driven by efficiency, this paradigm prioritizes people, blending physical, digital, and biological systems to promote sustainable progress. It advocates interconnectedness, combining technologies, values, and diverse stakeholders. In this vision, smart cities are not only efficient but also empathetic, designed to enhance the quality of life. Sectors like healthcare, education, governance, and industry converge, sharing data and insights to co-create solutions that resonate across society.

This transformative approach demands interdisciplinary collaboration and active citizen involvement, grounded in respect for local and global contexts. Digital inclusion, environmental care, cultural sensitivity, and social equity are cornerstones of its mission. Success in Society 5.0 is not about technical competence alone, but about its capacity to uphold human dignity and promote environmental sustainability. It calls for a paradigm shift from isolated solutions to interconnected systems, in which every innovation contributes to a thriving, thoughtful ecosystem that enhances all aspects of life.

1.6.1 Technological Pillars of an Interconnected Society

The technological foundation of Society 5.0 is built on systems that connect the cyber and physical worlds in real time. These technologies [7, 8] not only connect devices but also predict human needs. Artificial intelligence helps systems adapt, for example, by predicting disease outbreaks or improving energy planning. IoT and high-speed networks enable billions of sensors to share data, improving traffic flow, reducing fuel consumption, and supporting efficient farming. Big data and blockchain technologies securely manage large volumes of information, ensuring transparent supply chains and improved climate planning. Together, these technologies create a shared information system that supports timely, ethical, and proactive solutions.

1.6.2 Sustainability and Global Challenges

Society 5.0 embeds sustainability as a core part of national policy by integrating digital innovation with the UN Sustainable Development Goals (SDGs). Government plans and business rules guide both the public and private sectors in working together on clean energy, environmental protection, and social inclusion. Rules that use smart sensors strengthen environmental protection by monitoring ecosystems and efficiently managing renewable energy. Economic and health fairness is supported by the responsible use of data for pandemic planning, social support, and accessible services, while protecting privacy. Long-term planning includes hydrogen strategies that promote the use of clean energy across the power, industry, and transport sectors, thereby creating robust, interconnected systems.

1.6.3 Data, Trust, and Privacy in Interconnected Society

Data-intensive environments have the potential to significantly improve healthcare by supporting better diagnosis, treatment, and planning, but they face many practical and ethical challenges. Health data is often incomplete, unstructured, and scattered across hospitals, insurance systems, mobile apps, and personal devices, making it difficult to collect and integrate at scale. This process is also politically and socially sensitive, especially when data reveals patterns about vulnerable groups. Public concern is increasing around surveillance, data breaches, and the reuse of health data without explicit consent, particularly when such data is sold or used for commercial purposes. In response, governments and regulators are strengthening privacy laws, encouraging anonymization and pseudonymization, and enforcing stronger security measures, breach reporting, and oversight. Ethics committees, data protection authorities, and sector-specific regulators are gradually developing governance frameworks to manage data access, sharing, and algorithmic accountability. However, significant gaps remain. Consent forms are often difficult to understand; enforcement cannot keep pace with rapid technological change; and anonymized data can sometimes be re-identified. Citizens also have limited visibility into how their data influences decisions, leading to ongoing trust issues and uneven protection.

1.6.4 Eco-Burden of Data Boom

The rapid growth of data in the digital world has created serious environmental challenges. Everyday activities such as streaming videos, social media use, online searches, and AI queries generate enormous amounts of data that must be stored and managed. This data is handled by large data centres that consume vast amounts of electricity and require constant cooling, resulting in high carbon emissions. Globally, data centres already account for a significant share of total energy use, comparable to that of some countries. As Society 5.0 expands interconnected systems such as smart cities, digital healthcare, and IoT devices, the data volumes continue to rise. This growth increases energy demand, shortens hardware lifetimes, and increases electronic waste. Training a single large AI model further increases emissions; it can emit as much CO_2 as five cars over its lifetime.

These issues have led to growing awareness of what is often termed "digital pollution," with criticism directed at energy-intensive activities such as Bitcoin mining and large cloud platforms. In response, technology companies are investing in renewable energy, energy-efficient processors, and software that reduces unnecessary data storage and processing. Governments are also beginning to implement policies that monitor emissions from data centres and promote the use of low-carbon technologies. However, these efforts often struggle to keep pace with rapid growth in data. Long-term solutions require data pruning, improved efficiency, and a shift to edge computing. Combine this with carbon taxes on large technology firms and a stronger public push for greater transparency in footprints. If sustainability is embedded into digital systems from the beginning, Society 5.0 can balance innovation with environmental responsibility.

1.7 Conclusion

Society 5.0 moves beyond earlier stages of human development, such as hunting and gathering, agriculture, industrial production, and the information age. It emerged from Japan's efforts to address practical challenges faced by modern societies, including an aging population, growing healthcare needs, urban traffic congestion, environmental damage, and climate change. Rather than focusing solely on economic growth or technological advancement, Society 5.0 places people at the center of progress. It builds on lessons from history, recognizing that each stage of development improves life but also creates new challenges that must be addressed thoughtfully.

In this model, technologies such as AI, sensors, and large-scale data systems are tools to support human needs. For example, smart farming systems help reduce water use while increasing crop productivity, and digital health care tools enable early disease identification and better care. Urban systems can manage traffic, energy, and public services more efficiently, reducing stress on both people and the environment. These technologies are designed to work together smoothly, creating connected systems without confusion or overload.

A strong focus on people ensures that no group is left behind. Fair rules support changing job markets, education systems adapt to different learning needs, and services are designed to be easy to use. Secure data sharing connects individuals, communities, and infrastructure while respecting privacy and trust. Society 5.0 encourages cooperation among governments, businesses, and citizens to build greener cities, strengthen social connections, and create opportunities for all, aiming for progress that genuinely improves everyday life.

References

1. Deguchi, A., et al. (2020). What is Society 5.0? In *Society 5.0*. Springer. https://doi.org/10.1007/978-981-15-2989-4_1
2. Huang, S., Wang, B., Li, X., Zheng, P., Mourtzis, D., & Wang, L. (2022). Industry 5.0 and Society 5.0—Comparison, complementation, and co-evolution. *Journal of Manufacturing Systems, 64*, 424–428. https://doi.org/10.1016/j.jmsy.2022.07.010
3. Borra, S., Dey, N., & Peng, S. L. (2026). Human-centric AIoT strategies for sustainable business evolution. In *AIoT deployment in business evolution: Sociotechnical readiness and human-centric strategies* (pp. 47–57). Springer Nature Switzerland.
4. Bissadu, K. D., Sonko, S., & Hossain, G. (2024). Society 5.0 enabled agriculture: Drivers, enabling technologies, architectures, opportunities, and challenges. *Information Processing in Agriculture*. https://doi.org/10.1016/j.inpa.2024.04.003
5. Patil, R. V., Ambritta, N. P., Mahalle, P. N., & Dey, N. (2022). Medical cyber–physical systems in Society 5.0: Are we ready? *IEEE Transactions on Technology and Society, 3*(3), 189–198.
6. Khosravy, M., Gupta, N., Pasquali, A., Dey, N., Crespo, R. G., & Witkowski, O. (2023). Human-collaborative artificial intelligence along with social values in industry 5.0: A survey of the state-of-the-art. *IEEE Transactions on Cognitive and Developmental Systems, 16*(1), 165–176.
7. Mourtzis, D., Angelopoulos, J., & Panopoulos, N. (2023). The future of the human–machine interface (HMI) in Society 5.0. *Future Internet, 15*(5), 162. https://doi.org/10.3390/fi15050162
8. Borra, S., Dey, N., & Peng, S. L. (2026). *AIoT deployment in business evolution: Sociotechnical readiness and human-centric strategies*. Springer Nature Switzerland.

Chapter 2
Technological Infrastructure and Framework for Society 5.0

2.1 Introduction

Every generation of human civilization has been influenced by the infrastructure it has developed. Agricultural civilizations were established through fields and irrigation; thermal power and factories drove the industrial period; digital networks and worldwide connection advanced the information era. Society 5.0 is humanity's most ambitious, human-centric development to date. Society 5.0 fundamentally aims to integrate the cyber and physical universes. Artificial intelligence (AI), AI agents, robotics, the Internet of Things (IoT), blockchain, 5G and beyond, quantum computing, and big data analytics constitute the foundational elements of this emerging landscape. In contrast to other technical ages that predominantly emphasized productivity, Society 5.0 seeks to make it human-centric and to bring these technological advancements into the core of society [1]. It aims to establish ecosystems in which technology enhances human potential, addressing concerns such as aging populations, equity, climate change, accessibility, and resource constraints with intelligent and adaptive solutions.

Realizing this ambition requires not only technological advancements but also resilient, progressive infrastructure and a framework capable of managing complexity at scale. Society 5.0 necessitates seamless connectivity, interoperable platforms, and robust governance frameworks that foster trust and protect values, including dignity, equity, privacy, and security. In this perspective, infrastructure encompasses not only fiber-optic cables and sensor grids but also the ethical, cultural, and institutional frameworks that govern the deployment of technology and its beneficiaries.

The transition to Society 5.0 is both motivating and challenging. It provides the opportunity to reconceptualize education for continuous learning, to restructure health systems for anticipatory and personalized care, and to create cities that are sustainable, inclusive, and attuned to human needs. It also necessitates that we address possible disparities between those enabled by technology and those

S. Shukla et al., *Society 5.0*, SpringerBriefs in Applied Sciences and Technology, https://doi.org/10.1007/978-3-032-22164-3_2

marginalized, between innovation and fairness, between swift transformation and social preparedness. This chapter examines the technological infrastructure and systemic framework necessary to realize this objective. It advocates for infrastructure that is intricately interconnected, adaptive, and centered on human needs, capable of managing extensive data flows and intelligence, while aligning technological advancement with the enduring prosperity of society.

Society 5.0 proposes a future that is both deeply digital and human-centered. Its technological infrastructure, like quiet plumbing, lies beneath that vision; easy to ignore when it works well, impossible to miss when it fails. The main building blocks include universal sensing via IoT devices, AI for data analysis, cyber-physical systems that connect software with machines, ultra-fast 5G and 6G networks, blockchain-based trust platforms, emerging quantum computing, flexible cloud-edge setups, and layered cybersecurity. These components are not designed to function as separate "modules," but rather to work in concert to create an environment that facilitates the seamless flow of data from collection to interpretation, decision-making, and ultimately to some form of human-facing action.

Society 5.0's technological infrastructure is frequently referred to as its "central nervous system" in this regard. It detects environmental signals, attempts to interpret their significance, disseminates information through distributed networks, and assists in coordinating responses that promote sustainability, efficiency, and equity. In reality, how well this works rests less on any one part and more on how well they work together. This is consistent with the broader goal of integrating the real and virtual worlds, not merely in words.

Society 5.0 can be defined as a three-layered socio-technical architecture that combines the physical world, intelligent decision-making, and institutional trust into a single coherent system. The Physical-Cyber Interface Layer serves as the foundation, with sensors, gadgets, robots, and networks continuously sensing real-world circumstances and executing real-time actions. The Intelligence and Decision Layer is on top of it. This is the cognitive core that turns raw data into understanding through AI, advanced analytics, simulation, and optimization. This lets systems predict, reason, and help people make decisions at scale. The Trust, Governance and Resilience Layer, which encompasses both, guarantees security, transparency, ethical behavior, and societal legitimacy through cybersecurity, digital identity, and distributed trust mechanisms. These three layers work together to create a closed-loop, human-centered architecture in which technology actively augments human competence, resilience, and well-being, establishing Society 5.0 as a truly intelligent and trustworthy "Super-Smart Society".

2.2 Society 5.0 Framework

The Society 5.0 framework is a conceptual model intended to guide the transition from a society focused on Information (Society 4.0) to one that emphasizes human-centered technology use. The idea is to harness technology to build a human-first society. The basis of the idea is the integration of the cyber and physical worlds into an ecosystem in which information is continuously collected, evaluated, and transformed into useful knowledge that improves human life. Unlike earlier social models that relied largely on industrial efficiency, the Society 5.0 framework puts well-being, resilience, and inclusiveness in the forefront.

The approach recognizes that the growing global challenges, including population migration, climate crises, and rising socio-economic disparities, cannot be addressed through incremental changes. Rather, it requires a thorough assessment of the implementation and structure of technology within the societal framework. The main goal is to create a "super-smart society" that uses new technologies to meet people's needs, support sustainable growth, empower users, and improve community well-being.

2.2.1 Core Vision and Features

As depicted in Fig. 2.1, the Society 5.0 framework is multifaceted. It covers deeper aspects like human-centric approach, sustainability, and inclusivity, apart from enabling emerging technologies and smart infrastructure.

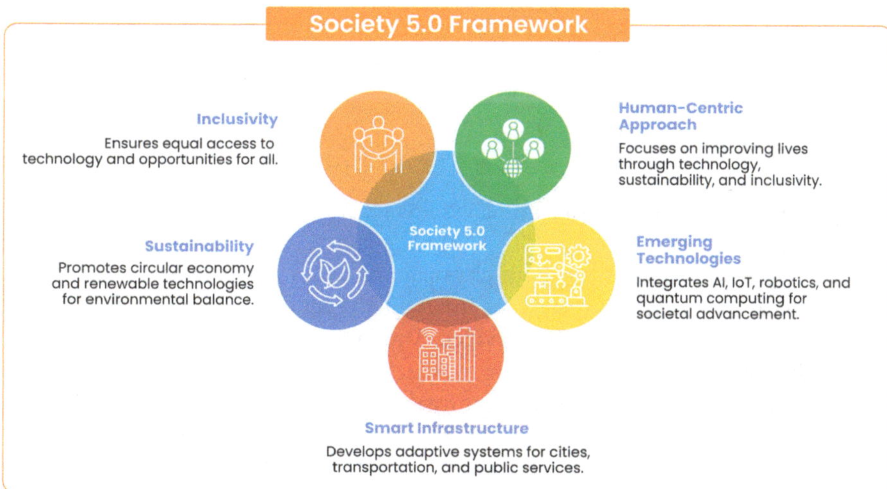

Fig. 2.1 Society 5.0 framework

2.2.1.1 Human-Centric Approach

In Society 5.0, the focus is on using the technology to improve people's lives, promote sustainability, and foster inclusivity through cutting-edge innovations. Imagine an AI-powered healthcare system that predicts, plans, and tailors treatment to improve patients' health. Web apps that use hyper-personalized services to meet users' very specific needs, taking personalization to a new level. A continuous education system that makes lifelong learning accessible through adaptive learning methods and easy-to-use digital tools. To enhance inclusivity, Robotics and assistive technologies can be used for older people, people with disabilities, and marginalized communities, enabling them to gain independence, mobility, and involvement in society.

2.2.1.2 Integration of Emerging Technologies

The framework relies heavily on a combination of advanced technologies, each with ground-breaking potential. The first is advanced Artificial Intelligence (AI), which enables AI agents to make decisions, predict outcomes, and operate autonomously. Then there is the Internet of Things (IoT), which provides continuous sensing and real-time data, giving autonomous systems eyes and ears. Robotics simplifies automation across manufacturing, healthcare, caregiving, and logistics. Big data analytics finds insights in the data that was previously too big to handle. Blockchain and distributed ledger ensure that transactions are open and trustworthy. Moreover, with the rise of Quantum Computing, we can now solve some computationally intensive problems that were previously considered infeasible. The amalgamation of these emerging technologies powers the Society 5.0 vision; they must work in tandem to achieve a truly human-centric society [2].

2.2.1.3 Smart Infrastructure and Systems

Society 5.0's infrastructure has both physical and cyber-physical parts. It is a huge network of infrastructure that includes smart cities, smart grids, intelligent transportation systems, and digital public services. The most important aspect of these systems is their adaptability. For example, a smart city can predict traffic jams before a political rally, reallocate grid energy in real time to handle festival peak load, or distribute healthcare resources more efficiently during emergencies such as COVID. These infrastructures are strong enough to withstand future crises and help societies become more efficient.

2.2.1.4 Sustainability and Circular Economy

The next era is about preserving Mother Earth; hence, Sustainability is the central idea of the Society 5.0 framework. We need to think about a circular economy that emphasizes conserving, recycling, and regenerating our limited resources. It is completely the opposite of previous societies, which focused solely on extracting, consuming, and disposing of resources. Society 5.0 promotes technologies such as renewable energy systems, smart farming, and smart water management that help preserve the environment and maintain ecosystem balance. To achieve this, we will use digital intelligence as a catalyst for sustainability. For example, smart cities can reduce carbon emissions by preventing traffic jams using predictive algorithms. They can also help improve crop growth with less water by using IoT-powered moisture sensors. IoT-enabled grids can also reduce energy waste in cities.

2.2.1.5 Inclusivity and Equity

If we want a smarter society, we need to address unfairness at its core. Hence, Society 5.0 emphasizes closing the digital divide. We need to ensure everyone has equal opportunities, regardless of socio-economic status. It means taking proactive steps, such as improving internet access in rural areas, making technology easier to use for people with disabilities, and starting reskilling programs for workers who lost their jobs due to automation. The idea is that technological progress should help most people, not just a few. Inclusivity is not optional; it is a key factor in the framework's effectiveness.

2.2.2 Key Components of the Framework

The dream of Society 5.0 is ambitious and transformative, but this cannot be realized until we put in place well-defined, structured components. Figure 2.2 shows key components of the Society 5.0 framework that would enable an easy transition of technology into societal activities. These components form the operational foundation of the framework, enabling the physical and cyber domains to work in tandem by exchanging data, insights, and decision-making. Each of the following components is a part of the foundational infrastructure required for a "super smart" society.

Society 5.0 begins with data collection as its foundation, in which sensors, wearables, drones, and edge devices continuously capture real-world information. Examples include smartwatches that monitor heart rates or soil sensors that alert farmers to irrigation needs. City traffic lights coordinate to reduce congestion, although peak hours reveal ongoing challenges. This data stream records current conditions and expects future developments, enabling adaptive governance when applied judiciously.

Data analytics processes this vast volume using AI and machine learning to extract actionable insights, such as predicting flu outbreaks across hospital networks or identifying policy correlations. Real-time analysis supports informed decisions and efficient resource allocation; yet, it requires careful consideration of trust and interpretation. Cyber-physical systems integrate these insights with physical actions,

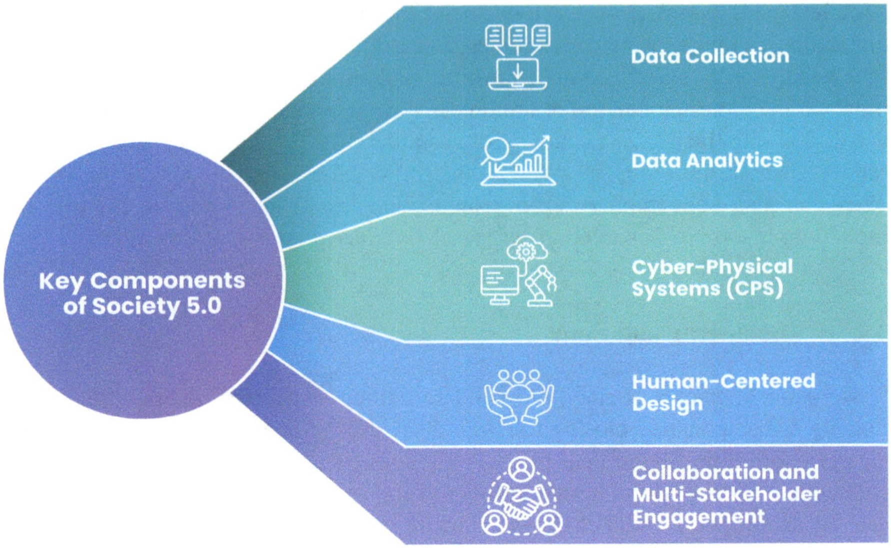

Fig. 2.2 Key components of Society 5.0 framework

as seen in self-driving cars or healthcare networks where human oversight remains essential.

Human-centered design prioritizes accessibility and inclusion, like wheelchair-friendly routes or voice-activated interfaces, over mere efficiency. Collaboration through Japan's quadruple helix model unites government, industry, academia, and citizens to drive cross-sector innovation and global alignment on ethics and standards. These components collectively foster an ecosystem balancing efficiency, equity, and resilience.

2.2.3 Framework Outcomes

The success of Society 5.0 is not measured merely by the number of technologies it deploys, but by what those technologies actually change for people, communities, and the planet. The framework aims to reshape how societies function, blending the cyber and physical worlds in ways that feel human rather than mechanical.

Figure 2.3 summarizes the outcomes of the Society 5.0 framework. Efficiency, sustainability, and fairness can be the primary and most visible metrics for assessing progress. There are additional nuances, such as how well the system builds resilience, nurtures the capacity to adapt, and, finally, how it encourages purposeful collaboration across cultures and regions. These outcomes depend heavily on how well the system is implemented, or more specifically, on whether people trust the system designed to benefit them.

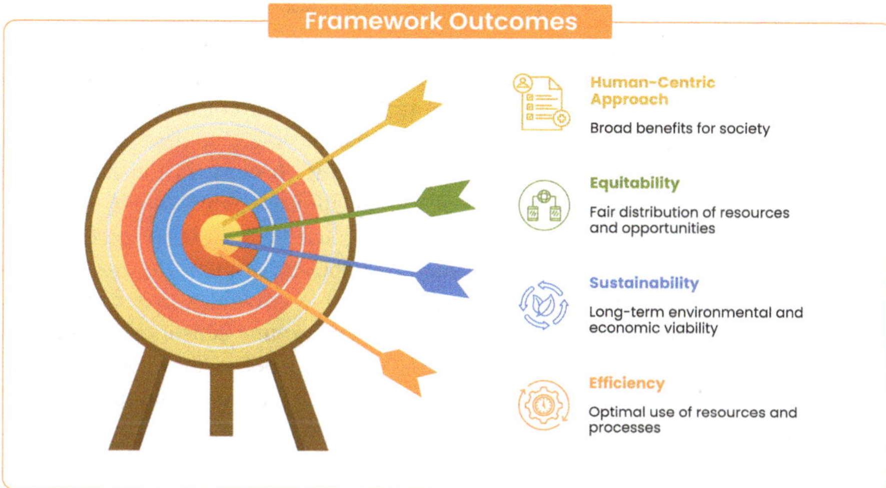

Fig. 2.3 Framework outcomes

2.2.3.1 Efficiency

One of the important goals of Society 5.0 is Efficiency. The very idea of ensuring that technology helps us do things not only faster and smarter, but with fewer wasted resources. Moreover, the goal of efficiency is not just speeding things up, but doing so smoothly and quietly that we sometimes barely notice their seamless operation. For example, in healthcare, predictive models can flag early signs of illness. This helps doctors intervene before hospital visits become critical. Wearable health monitors, once seen as gimmicks, are increasingly assuming real responsibilities, tracking heart rhythms, oxygen levels, and even stress patterns. Traffic management tells a similar story. Cities experimenting with IoT-enabled roads adjust signal timings in real time, easing morning bottlenecks and cutting down emissions that most commuters have stopped paying attention. Agriculture, too, has seen this shift. Farmers using sensor-guided irrigation might decide when to water not by intuition but by data from buried moisture probes. It doesn't eliminate human judgment, but it changes how that judgment is made; less guesswork, fewer wasted inputs, and often better yields. We do not yet know whether such models can scale sustainably, but they point to a future in which efficiency feels more like daily ease than optimization.

2.2.3.2 Sustainability

The goals of Society 5.0 are closely aligned with those of the United Nations Sustainable Development Goals (SDGs) and other global environmental initiatives. But it is not just about checking items off a list; it is about changing the way societies make, use, and preserve things.

Energy is usually the first thing. Smart grids and renewable energy sources, such as solar, wind, and local storage systems, are transforming how electricity flows through cities and towns [3]. These systems do not just replace fossil fuels; they change how people use energy in general, turning passive consumers into active participants in a grid that learns and adapts. The same logic applies to materials. IoT-based tracking enables circular economy models to work in the real world. They help you trace where resources go after they are used, which helps you identify problems and develop new ways to recycle or reuse them. It is not perfect; sensors fail, data gaps show up, and sometimes the economics do not support sustainability. But the goal is a step toward responsible growth. Another promising area is environmental monitoring. Now, networks of ground sensors and satellites work together to warn people about floods or earthquakes before they happen. These systems provide communities with a few additional minutes or hours to get ready, which can make a big difference, even though they are not always accurate.

2.2.3.3 Equitability

One of Society 5.0's most daring promises is that it will work for fairness, meaning that technology should close, not widen, the gaps that separate people [4]. It is a substantial claim, and in practice, it is difficult to maintain. The framework imagines tools and networks that are not just for a select few but for everyone, no matter where they live or what resources they have.

For example, high-speed 5G and satellite-based internet are slowly changing how people get education and health care in rural areas. A child living in a remote mountain village might now be able to take virtual classes on digital learning platforms that were once available only to urban schools. An interesting example of a demographic problem is Japan's aging population. In this scenario, equity can be achieved through robots that assist with social tasks, telehealth consultations, and AI-supported healthcare tracking, thereby making it easier for the target population to live independently. Rapid automation can also make systems less stable. For example, rapid AI adoption is making the job market unstable, where people are required to quickly adapt to rapid technological change or perish. Some countries are addressing this by offering vocational training and targeted reskilling programs. Moreover, these programs are AI-powered to suit an individual's strengths. Although the effectiveness of such a program has yet to be seen in practice, it is a step in the right direction. It is a path guided by the uneasy yet necessary promise of fairness in technology.

2.2.3.4 Societal Outcomes

Society 5.0 extends beyond efficiency, sustainability, and equity to deliver broader strategic benefits that strengthen communities and systems. Resilience is a key measure of success. Cyber-physical monitoring enables faster responses to crises like pandemics, cyberattacks, and natural disasters such as floods and wildfires. True resilience, however, goes further. It demands proactive planning, reliable backups, and community trust, elements that technology supports but cannot create alone. Innovation ecosystems thrive through collaboration among government, academia, industry, and citizens. This co-creation approach has helped Japan strengthen its research capacity and enhance global competitiveness. Yet, it requires cultural shifts, especially in hierarchical settings where openness does not come naturally. Global cooperation pushes the vision forward. Initiatives like "Society 5.0 for SDGs" promote shared standards, ethical technology exchange, and aligned governance. Challenges persist, including those related to national interests and resource gaps. Ultimately, success rests on how societies wield these tools to enhance lives, economies, environments, and fairness.

2.3 Physical-Cyber Interface Layer

Society 5.0 hinges on a seamless bridge between the physical world and digital systems, referred to as the physical–cyber interface layer. Here, technologies like IoT, CPS, and next-generation networks work together to enable continuous, real-time awareness and action. IoT forms the sensory base, quietly gathering data from wearables to soil sensors; CPS turns AI insights into physical moves like adjusting traffic lights; while 5G/6G ensures all flows without lag. Japan's pilots, from Tsukuba's smart monitoring to Fukushima's drones, test this fusion in practice, aiming for efficiency, equity, and resilience, though full convergence may still face hurdles like scalability.

2.3.1 Internet of Things (IoT): The Sensory Layer of Society 5.0

2.3.1.1 Framework Context

In the Society 5.0 vision, the Internet of Things acts like the sensory layer of a "super-smart" society. It quietly links sensors, devices, and small intelligent nodes across homes, factories, public spaces, and even natural environments, so that the digital and physical sides of the system can stay connected in a constant and near real-time manner [5]. Here, IoT forms the data-gathering layer on which forecasting, algorithmic decision-making, and adaptive responses are likely to depend.

2.3.1.2 Core Components of IoT in Society 5.0

- Comprehensive Sensing: Wearable medical devices that track heart rate or blood oxygen saturation throughout the day, or soil sensors that alert farmers when fields are running dry, are one of the many concrete examples of how IoT captures signals from both human activity and the environment.
- Interoperable Platforms: Shared standards and communication protocols enable devices from different vendors to "talk" to each other, which is crucial for dismantling incompatible silos and achieving scalability in smart systems.
- Edge Devices and Gateways: More computation is gradually moving to devices themselves or nearby gateways, which can cut latency and reliance on distant cloud servers when a quick response really matters.
- Integration with Intelligent Systems: These networks integrate with smart-city dashboards, traffic-control systems, energy grids, and healthcare platforms, so data does not remain in logs but shapes day-to-day decisions.

2.3.1.3 Outcomes

- Efficiency: Continuous monitoring of industrial equipment can flag problems early, cutting unplanned downtime and smoothing production, while smart meters and grids help adjust electricity and water use to match real demand.
- Sustainability: IoT-enabled precision agriculture can reduce unnecessary irrigation and fertilizer use, and connected energy systems can better align consumption with locally generated renewable energy.
- Equity: Remote diagnostics and connected health devices enable access to basic medical care for people living far from large hospitals or who find frequent in-person visits difficult, such as older, and people with disability.

In Japan's Tsukuba Smart City initiative, IoT is used for traffic monitoring, optimizing waste-collection routes, tracking air quality, and turning the city into a kind of living laboratory for Society 5.0 ideas. IoT-based telemedicine pilots that link specialists in urban hospitals with patients in outlying regions hint at how this "sensory layer" might support not only efficiency, but a more inclusive service delivery.

2.3.2 Cyber-Physical Systems (CPSs): The Integrative Nexus of Society 5.0

2.3.2.1 Framework Context

Within Society 5.0, Cyber-Physical Systems connect code with the real world. IoT devices collect signals from the real-world; AI tries to interpret them; and CPS is where those insights translate into real-world actions, such as traffic lights changing, machines adjusting, and robots moving. In this way, CPS operates as a "dynamic interface," where decisions and physical behavior are intended to remain in constant, near-real-time synchronization. By tightly weaving together computation, networking, and physical processes, CPS supports a two-way loop: cyberspace proposes interventions that manifest as actions in the physical world. In contrast, the physical environment continually feeds data back to refine the digital models. This continuous cycle sustains the Society 5.0 machinery [6].

2.3.2.2 Core Components of CPS in Society 5.0

- Sensor-Actuator Integration: IoT networks stream data from sensors, while actuators, such as robotic arms, automated conveyor belts, traffic signals, or autonomous shuttles, carry out responses based on AI-generated decisions.
- Architecture: Instead of treating sectors such as energy, mobility, healthcare, and manufacturing as separate islands, CPS ties them together through shared platforms, enabling cross-sector monitoring and coordination.
- Real-Time Synchronization: CPS aims to enable systems such as power grids and transport networks to adjust instantly to shifting demand, equipment failures, or unexpected events, reducing delays and nasty surprises.
- Autonomy with Human Oversight: Many CPS setups lean toward machine autonomy. Self-driving vehicles are the obvious example, but they are supposed to operate under human-defined constraints, ethical rules, and safety protocols, with humans still able to intervene or override.
- Resilience and Adaptability: These systems are designed to detect anomalies, reroute around failures, and reconfigure after disruptions; when implemented effectively, they can improve the long-term durability of critical infrastructure.

2.3.2.3 Outcomes of AI in Society 5.0

- Efficiency: CPS-enabled Smart factories can enhance production efficiency by cutting down on downtime. They can also improve supply chain efficiency by reducing delays in the movement of goods within supply chain systems. It can be replicated beyond and applied to autonomous delivery systems and warehouse robots.
- Sustainability: Data-driven solar panels can utilize CPS to coordinate with rooftop solar panels and neighborhood storage to identify the demand-supply gaps. They can further augment the data with weather patterns and changing demand, making it much easier to extract the best possible renewable energy without disrupting the power grid.
- Equity: One of the most effective applications of CPS is in healthcare, where CPS systems connect wearables, remote diagnostic tools, and assistive robots. It is intended to ensure that people in remote areas or aging communities receive periodic checkups and timely care in emergencies.

- Resilience: Digital systems are prone to faults. CPS can support early identification of hazards, such as predictive mechanical failures, structural fatigue in smart buildings, air-quality and toxicity spikes, public health threats, and intrusion & spoofing, to identify cyber-terrorists, etc. CPS can identify early and frame targeted responses through predictive analytics embedded in daily operations.

In Japan, Autonomous vehicle trials combine dense sensor networks, AI planning systems, and real-time control to adapt routes and speed to traffic, weather, and passenger needs, rather than following rigid schedules. Smart energy projects pair household solar panels with neighborhood-level controls to keep local supply and demand in balance. Robot-assisted surgery and AI-supported imaging in hospitals demonstrate how cyber-physical coordination can enhance system security and reliability. At the same time, they still depend heavily on human judgment at critical moments.

2.3.3 5G/6G Networks: The Communication System of Society 5.0

2.3.3.1 Framework Context

If IoT gives Society 5.0 its senses, AI its thinking, and CPS its ability to act in the real world, then 5G and, eventually, 6G are the arteries and veins that keep everything moving. These networks carry data, model outputs, and services across homes, streets, factories, and public institutions, often so quietly that most people barely notice, until the connection drops. The idea is that this "communication system" keeps the whole socio-technical body alive and responsive. To make that vision work, the framework relies heavily on ultra-low latency, high-bandwidth, and extremely reliable links. Without that mix, real-time coordination among billions of devices, applications, and human users quickly breaks down. When connectivity is patchy or slow, the promised convergence of the cyber and physical worlds tends to look less like a seamless experience and more like a series of frustrating lags and blind spots, which is exactly what Society 5.0 aims to move beyond.

2.3.3.2 Core Components of Next-Gen Networks in Society 5.0

- High Bandwidth and Speed (5G and beyond): Fast, high-capacity links are what make data-heavy scenarios feel normal rather than futuristic, think high-definition telemedicine calls where a specialist in metro city examines scans for a patient in a rural clinic, or autonomous vehicles constantly exchanging sensor data with roadside units. They also underpin real-time AR/VR in classrooms or training centers, where even a 1-s lag can disrupt the experience.

- Ultra-Low Latency (~1 ms in 5G, aiming for near-zero in 6G): Some operations simply cannot tolerate delay. Remote surgery, automated disaster-response systems, or tightly coordinated industrial robots all rely on near-instant feedback. Even a small spike in latency may translate into a missed command or a safety risk in these cases.
- Massive Machine-Type Communication (mMTC): Next-gen networks are built to handle not just thousands of users but billions of devices chirping away in the background, e.g., streetlights, soil sensors, wearables, factory machines. The goal is a sensor-rich environment where this chatter does not clog the network whenever a city rolls out a new "Smart" service.
- Network Slicing: Instead of treating connectivity as one big, undifferentiated pipe, slicing lets operators carve the network into virtual lanes. An emergency service channel, for instance, can guarantee bandwidth and reliability even when everyone else is streaming sports highlights, while a less critical slice handles ordinary consumer traffic.
- Integration of 6G Features: Looking toward the 2030s, 6G concepts go a step further. They imagine networks with AI built into services across layers, stronger security through techniques like quantum key distribution, terahertz-band links for extreme data rates, and tight integration with satellite and other space-based systems. While 5G connected people and things, 6G is designed to connect the "digital, physical, and humane worlds."

2.3.3.3 Outcomes of 5G/6G in Society 5.0 Framework

- Efficiency: High-speed, low-latency networks let logistics chains operate more like coordinated systems than disconnected steps. Fleets, warehouses, and ports can share status in real time, while remote sensors push equipment data directly into predictive maintenance systems, eliminating the need for monthly inspections.
- Sustainability: When connectivity is reliable, smart grids can integrate solar, wind, storage, and demand response at scale rather than treating renewables as awkward add-ons. This fine-grained balancing helps avoid blackouts and reduces the need for carbon-intensive backup generation.
- Equity: Fast, reliable networks are also what make serious telemedicine and online education plausible beyond big cities. A student in a remote village attending a live interactive class, or an older patient having a video consultation with a specialist, depends as much on uplink stability as on any app design. Without that, "access" remains mostly a promise on paper.
- Resilience: During earthquakes, typhoons, or other large-scale disruptions, prioritized network slices can keep emergency communications, hospital links, and command centers functioning even when the rest of the system is under stress. Drones equipped with cameras and sensors can be dispatched over 5G networks to map damage or search for people in hard-to-reach areas on foot.

Japan offers some concrete glimpses of where this may be heading. In Fukushima, 5G-connected drones have been used in disaster-recovery exercises to stream live aerial footage back to responders. Pilot projects in Tokyo have tested remote consultations and elements of robot-assisted surgery over high-speed networks. At the same time, autonomous bus trials adjust routes and frequency based on actual passenger loads and road conditions rather than fixed schedules. Looking ahead, both Japan and South Korea are investing heavily in 6G research, treating it less as a simple speed upgrade and more as "cognitive connectivity," where the network itself hosts AI that can observe, learn, and adapt alongside the services it carries.

2.4 Intelligence and Decision Layer

Society 5.0's intelligence layer transforms raw data into actionable outcomes, with AI as the cognitive core that filters patterns from IoT streams into human-relevant insights. Quantum computing accelerates the toughest problems, like genomic modeling or climate forecasts, while cloud-edge architectures balance split-second local decisions with global-scale analysis. Together, these elements form a flexible "brain" for the system. AI for everyday reasoning, quantum for breakthroughs, and hybrid computing for adaptability across edge devices and vast data centers.

2.4.1 Artificial Intelligence (AI): The Cognitive Engine of Society 5.0

2.4.1.1 Framework Context

If IoT gives Society 5.0 its senses, AI is meant to serve as its thinking brain. It takes in raw, messy data streams from millions of connected devices and turns them into patterns, suggestions, and sometimes fully automated actions. In this framework, AI is not treated as just another technical add-on, but as a strategic layer that shapes human-centered decisions, public policy, and the way systems adapt over time. The basic requirement is that these large data ecosystems become legible and socially useful, shaping outcomes toward well-being, inclusion, and sustainability rather than mere efficiency.

2.4.1.2 Core Components of AI in Society 5.0

- Machine Learning and Predictive Analytics: Models pick up patterns, estimate what is likely to happen next, and enable preventive action, such as flagging early signs of disease in health records or predicting energy demand before a heatwave hits.

- Natural Language Processing and Human–Machine Interfaces: Tools like conversational agents or voice-based assistants lower the barrier to using complex systems, especially for people who are not comfortable with technical interfaces, for example, elderly individuals managing medication reminders through speech rather than apps and menus.
- Computer Vision and Robotics: Specialized AI systems understanding visual and spatial data to command automated machinery and robots in various domains, including self-driving cars, fulfilment center sorting robots in logistics hubs, and even with medical imaging, where even minor anomalies in scans might be too small for human eyes to catch.
- Cognitive Decision Systems: In the field of urban control centers, emergency response platforms, and policy dashboards, AI can increasingly be crucial. It can be used to weigh options, simulate scenarios, and surface trade-offs, ideally with some awareness of ethics, regulations, and local context, rather than relying solely on numerical optimization.
- Responsible and Ethical AI: The key for Society 5.0 is that AI be transparent, accountable, and aligned with social values. That includes, but is not limited to, attention to privacy, bias, explainability, and the simple question of whether people can contest or understand AI-assisted decisions.

2.4.1.3 Outcomes of AI in Society 5.0

- Efficiency: AI can reduce redundancies, anticipate failures in infrastructure, and adjust operations on the fly, for instance, public transport routes that shift with real-time demand instead of following rigid timetables.
- Sustainability: Models that fine-tune heating and cooling in buildings, guide precision farming, or forecast floods and landslides may help cut waste and environmental damage, even if the predictions are not perfect.
- Equity: In theory, artificial intelligence-powered applications such as personalized digital tutors or cost-effective smartphone diagnostic apps could help reduce disparities in access to education and healthcare, especially for people living far from major institutions or who do not receive sufficient services.
- Resilience: By looking at big, mixed datasets like hospital records, mobility data, and network logs. AI can help identify new problems sooner, such as pandemics, cyberattacks, or shocks caused by climate change.

Japan's own experiments suggest what this looks like in practice. AI-driven imaging systems help clinicians detect diseases earlier. At the same time, predictive models tailor care plans for older adults rather than treating "the elderly" as a single, uniform group. Similar ideas show up in agriculture, where AI supports smart farming; in mobility, through work on autonomous vehicles; and in urban management, where traffic flows and disaster responses are continuously tweaked. Together, these efforts position AI as a kind of core component of decision-making for Society 5.0, though how well it actually balances technological ambition with broader social concerns remains an open question.

2.4.2 *Quantum Computing: The Acceleration Engine of Society 5.0*

2.4.2.1 Framework Context

Artificial intelligence, the Internet of Things, cyber-physical systems, and blockchain all entail significant computational burdens. Classical computing has done the groundwork for the Society 5.0 idea. Still, the kinds of problems arising now include high-resolution climate models, genomic medicine, global logistics, and large-scale financial optimization. They are starting to stretch conventional architectures to their limits. Quantum computing enters this picture less as a shiny standalone trophy and more as an accelerator: a way to tackle classes of problems that were previously considered as practically impossible. It is meant to amplify what is already there, letting AI handle even larger and more complex datasets, strengthening cryptography behind shared ledgers, and supporting cyber-physical operations across entire cities or even national infrastructures [7].

2.4.2.2 Core Components of Quantum Computing in Society 5.0

- Quantum Algorithms: Designed to exploit quantum parallelism for tasks like optimization, cryptography, and certain machine-learning workloads, where small speedups can translate into big real-world gains.
- Quantum–AI Integration: AI pipelines augmented with access to quantum backends may be able to cope with exponential data volumes in areas such as genomics, detailed weather and climate simulations, or highly complex scenario modeling.
- Quantum Cryptography: Approaches such as Quantum Key Distribution (QKD) aim to secure communications against attackers who might one day have quantum capabilities, supporting the long-term security of critical systems.
- Hybrid Computing Architectures: In the near term, classical cloud infrastructure is expected to work alongside smaller quantum accelerators, with most tasks staying classical. At the same time, specific sub-problems get offloaded to quantum hardware.
- Domain-Specific Applications:
  - Health: speeding up drug discovery pipelines and enabling fine-grained personalized treatment planning.
  - Energy: exploring new materials for better solar cells, batteries, and hydrogen storage, and improving optimization in smart grids.

- Climate: running more detailed environmental models to support earlier and more targeted mitigation strategies.

2.4.2.3 Outcomes of Quantum Computing in Society 5.0 Framework

- Efficiency: Routing for global logistics, complex supply-chain optimization, or certain financial computations could be solved much faster than on traditional systems, leading to sharper, data-driven resource allocation.
- Sustainability: By accelerating the search for advanced materials and improving large-scale system optimization, quantum tools may expedite the transition to cleaner energy technologies.
- Equity: While early quantum hardware is likely to be expensive and concentrated, cloud-style access models might lower the barrier over time, allowing smaller countries, firms, and universities to tap into high-end computation instead of being permanently locked out.
- Resilience: Quantum-resistant cryptography and quantum-enhanced security techniques are intended to protect vital digital infrastructure against future attackers equipped with powerful machines of their own.

Japan has begun framing quantum technology as a key building block of its Society 5.0 trajectory, with programs such as the Moonshot R&D initiative funding research into quantum computing for materials science and drug development. Collaborative projects linking universities, government labs, and major industrial players focus on applications in manufacturing optimization and healthcare. At the same time, related efforts in the United States and Europe suggest a broader, cross-border ecosystem emerging around similar goals. Taken together, these moves point to a future in which quantum computing is less a separate "quantum era" and more a deeply embedded layer in the wider Society 5.0 landscape, quietly powering some of its most demanding tasks.

2.4.3 Cloud–Edge Architectures: The Adaptive Nervous System of Society 5.0

2.4.3.1 Framework Context

In the Society 5.0 vision, "intelligence" is meant to be something you can tap into almost anywhere: on the street, at home, in a clinic, or in a factory, through billions of connected devices. To achieve that, you need more than just fast chips; you need a computing backbone that can stretch and adapt without failing. That is where cloud–edge designs come in, blending the raw muscle of big data centers with the closeness and quick reactions of local processing nodes. The edge handles what needs to happen right here, right now, while the cloud handles the heavy thinking and long-term memory.

Together, this creates an adaptive nervous system for Society 5.0: the edge offers proximity, low latency, and a bit of fault tolerance, and the cloud brings analytical depth and vast storage. When it works well, intelligence is not locked away in a few elite facilities but spread across communities, cities, and even individual devices. That distribution ties back to the three big promises of Society 5.0, including greater efficiency, more sustainable use of resources, and fairer access to digital services.

2.4.3.2 Core Components of Cloud–Edge Computing in Society 5.0

- **Cloud Platforms**: Centralized environments where massive datasets are aggregated, CPS are coordinated, and AI models are trained and updated before being pushed back out to the edge.
- **Edge Computing Nodes**: Small but capable processors embedded in smart grids, cars, medical equipment, and IoT devices that can run real-time analytics without waiting for a round trip to a distant data center.
- **Edge–Cloud Synergy**: In this, the workloads are splited as time-critical tasks, such as obstacle detection in autonomous vehicles or anomaly spotting in a factory, and stay at the edge. At the same time, model retraining or long-range forecasting, compute-intensive jobs are handled in the cloud.
- **Fog Computing and Orchestration**: Intermediate layers coordinate swarms of edge devices, manage updates, and smooth out scaling issues so that thousands of distributed nodes still behave like a coherent system rather than a collection of gadgets.
- **Data Governance and Privacy**: Processing data as close as possible to its source reduces the need to ship sensitive information around, which fits the human-centered promise of Society 5.0 and lowers privacy risk.

2.4.3.3 Outcomes in the Society 5.0 Framework

- **Efficiency**: Millisecond-level responses at the edge support things like autonomous transport, drone delivery, and on-the-spot predictive maintenance, while centralized cloud systems look across many sites to optimize routes, schedules, and resource use at a broader scale.
- **Sustainability**: Local processing reduces the constant data back-and-forth with centralized servers, trimming energy use and network overhead, especially when sensors stream information continuously.
- **Equity**: Cloud–edge hybrids can bring AI-powered services like health monitoring, digital learning, and advisory tools to smaller towns and rural areas, not just dense urban centers with top-tier infrastructure.

- **Resilience**: If a central service goes down or a disaster disrupts backbone connectivity, edge nodes can still keep essential functions running locally, which gives communities a better chance of coping under stress.

Japan's investments in cloud–edge smart city projects, such as the Kashiwanoha initiative, offer a glimpse of how this might look in practice. In healthcare, edge-enabled wearables continuously track vital signs, while cloud systems aggregate and analyze the data to generate broader public-health insights. In mobility, autonomous buses process sensor data on board for moment-to-moment navigation, while city-level traffic coordination runs in the cloud. For energy, household-level edge nodes fine-tune consumption in response to local conditions, even as cloud platforms manage balancing across the wider grid. Taken together, this kind of architecture shows how distributed intelligence can stay relatively human-centric by respecting context, privacy, and local needs, while still aligning with the larger ambitions of Society 5.0.

2.5 Trust, Governance, and Resilience Layer

Society 5.0's trust layer addresses the hidden risks of hyper-connectivity, where blockchain builds verifiable data trails, and cybersecurity acts as a quiet immune system against breaches. Blockchain fosters transparency in transactions from energy trading to health records, while cybersecurity enforces zero-trust, privacy-by-design, and resilient defenses across IoT, AI, and CPS ecosystems. These elements ensure that "Super-Smart" society serves people, not just systems, by prioritizing accountability, equity for underserved groups, and recovery from disruptions like cyberattacks or disasters.

2.5.1 Blockchain and Distributed Ledgers: The Trust Architecture of Society 5.0

2.5.1.1 Framework Context

In the Society 5.0 vision, the widespread adoption of IoT, AI, CPS, and next-generation networks means that almost everything generates or processes data. That, in turn, raises some uncomfortable but necessary questions: who actually owns this data, who gets to see it, and under what conditions is it shared or locked. In a hyper-connected environment where billions of devices are communicating, and countless transactions unfold at the same time, blockchain and other distributed ledger technologies are often presented as a way to anchor trust, offering transparency, security, and a record that is hard to rewrite quietly.

Within this framework, blockchain systems are imagined less as crypto engines and more as basic social infrastructure. They help verify transactions, keep shared datasets from being silently altered, and make it harder for any single actor to control the narrative. By spreading verification across a network instead of concentrating it in one institution, these systems aim to boost citizen trust, strengthen institutional credibility, and create fairer conditions for participation.

2.5.1.2 Core Components of Blockchain in Society 5.0

- Decentralized Ledgers: Distributed records of transactions or data exchanges that are designed to be tamper-evident and publicly auditable, making quiet back-room edits much harder [8].
- Smart Contracts: Self-executing agreements that trigger automatically when certain conditions are met. It is useful for things like supply-chain payments, public procurement rules, or automated subsidies, where manual enforcement is slow or patchy.
- Interoperable Chains and Platforms: Mechanisms that let different blockchain systems to talk to one another, helping avoid a future of fragmented islands of data that cannot be shared or verified across sectors.
- Identity and Data Sovereignty: Digital identity frameworks that let individuals and organizations prove who they are while still retaining some control over which parts of their data they reveal, and to whom.
- Integration with IoT and AI: Sensor data can be signed and logged on a ledger, making its origin and integrity easier to verify, while AI systems that rely on such data can point back to tamper-resistant trails when their decisions are questioned.

2.5.1.3 Outcomes of Blockchain in Society 5.0

- Efficiency: In manufacturing and agriculture, end-to-end tracking enables companies and regulators to trace goods from source to shelf, reducing disputes, fraud, and the need for manual reconciliation.
- Sustainability: Documenting material flows on shared ledgers can support responsible sourcing, trace recycled inputs, and give circular-economy claims a firmer foundation than marketing language.
- Equity: Secure, portable digital identities may open doors to banking, welfare schemes, or healthcare for people who lack conventional paperwork, including Older people, residents of remote areas, or migrant workers.
- Resilience: For critical systems such as energy or healthcare, using ledgers to secure key records can make it harder for attackers or even internal actors to tamper with data precisely at the moment it is most needed.

Japan has been experimenting along these lines in several domains. Pilot platforms for peer-to-peer renewable energy trading, for instance, let households with rooftop solar sell excess power directly to neighbors, with each transaction logged on a shared ledger. In healthcare, blockchain-secured records give patients greater control over who accesses their data, while still allowing hospitals and clinics to share information when it matters. Smart agriculture projects track produce from farm to supermarket, giving consumers a clearer picture of its origin and handling, and offering farmers a verifiable account of their practices. Collectively, these examples echo a core idea of Society 5.0: technology should not only squeeze out more efficiency, but also foster trust, support accountability, and give people a stronger sense of control over the systems that shape their lives.

2.5.2 Cybersecurity: The Immune System of Society 5.0

2.5.2.1 Framework Context

In Society 5.0, cybersecurity emerges as the immune system protecting a hyperconnected ecosystem where IoT senses, AI decides, CPS acts, and data flows constantly between physical and digital worlds. Traditional perimeter defenses fall short against threats that could cascade from a single compromised sensor to entire hospitals or power grids. Instead, this approach demands adaptive layers such as zero-trust verification, AI-driven anomaly detection, and privacy woven into every process to sustain trust amid humans, machines, and autonomous systems sharing the same vulnerable space. Without it, real-time systems could easily break.

2.5.2.2 Core Components of Cybersecurity in Society 5.0

- **Zero-Trust Architectures**: Instead of assuming anything inside a network is “safe,” zero trust treats every user, device, and request as potentially compromised. Continuous authentication and verification help limit damage from stolen credentials, insider mistakes, or a single infected IoT sensor quietly sitting in a factory or home network.
- **Ai-Enhanced Threat Detection**: Machine learning models review logs, network flows, and device telemetry to detect anomalous patterns, including connections at unusual hours, unusual data transfers, or behavior that does not conform to past norms. This kind of adaptive monitoring appears better suited to fast-moving threats than rule sets alone, although it also raises questions about false positives and over-reliance on automated decisions.
- **Quantum-Resistant Cryptography**: As quantum computing progresses, some of today’s widely used cryptographic schemes may become vulnerable. Preparing for this shift means gradually introducing algorithms expected to withstand attacks, even from quantum-capable adversaries, especially around critical infrastructure and long-lived sensitive data.

- **Privacy by Design**: Rather than tacking privacy on the end, systems are meant to include minimization, consent, and clear data flows from the start, across IoT devices, CPS platforms, ledgers, and AI services. The aim is not only legal compliance but also a measure of dignity and autonomy for people whose data powers these systems.
- **Resilient Infrastructures**: Redundancy, segmentation, backups, and safety modes help essential services like power, transport, hospitals, and emergency coordination to continue functioning even when parts of the network are under attack or have failed. Here, the focus shifts from "never be breached" to "keep operating and recover quickly."
- **Human Factor Awareness**: People are still the most common entry point for attackers, but they can also be the strongest line of defense. Training, clear procedures, and a culture that encourages reporting suspicious activity turn citizens and employees from "weak links" into active partners in security, which fits the human-centered language of Society 5.0.

2.5.2.3 Outcomes of Cybersecurity in the Society 5.0 Framework

- **Efficiency**: When systems are perceived as trustworthy, individuals and organizations are more willing to embrace automation, whether that is stepping into an autonomous shuttle, authorizing digital payments, or letting software manage parts of an industrial process, without constant hesitation.
- **Sustainability**: Data-heavy infrastructure, such as smart grids or connected water systems, relies on accurate, uncompromised information to function. Protecting them from tampering or disruption helps keep renewable energy and other sustainable services reliable rather than fragile.
- **Equity**: Secure digital access matters most to people with the fewest alternatives. If online services are unsafe, it is often marginalized groups, those with limited mobility, patchy local services, or fewer resources, who are affected the most. Strong cybersecurity, paired with inclusive design, can reduce that uneven exposure.
- **Resilience**: A society that expects cyber incidents, plans for them, and rehearses its responses is better positioned to ride out attacks, natural disasters, and cascading failures. Maintaining public trust in digital systems under stress may be one of the quiet but critical tests of Society 5.0.

Japan's national strategy explicitly frames cybersecurity as a core enabler of its broader Society 5.0 agenda, linking incident readiness and policy coordination with the roll-out of smart city pilots and digital public services. Urban testbeds have begun adopting zero-trust approaches and AI-based anomaly detection to keep essential services running even under attempted intrusions, while sectors like healthcare experiment with combinations of secure records and privacy-preserving analytics to protect sensitive patient data without blocking interoperability. Similar debates and regulatory moves in regions such as the European Union, the United

States, and China suggest a growing consensus that advanced "intelligent" societies cannot function without deeply integrated security and governance, even if each region interprets that balance differently.

2.6 Conclusion

This chapter shows how Society 5.0 emerges less from a single breakthrough technology than from the careful weaving together of several key infrastructures into something that feels cohesive. The IoT provides the sensory layer, letting societies maintain a constant, real-time pulse on themselves. Artificial intelligence steps in as the thinking part, turning floods of raw data into something resembling understanding. Cyber-Physical Systems bridge the gap between digital plans and physical action, while 5G and 6G networks act like the bloodstream, rushing intelligence where it is needed with the speed that makes immediacy possible.

Still, all that responsiveness means little without some foundation of trust. Blockchain and distributed ledgers offer a way to verify data origins, ensure transparency, and build public faith in these digital exchanges, though whether they fully deliver on that promise in messy real-world settings remains an open question. Quantum computing tackles the complex problems that classical systems struggle with, from drug design to climate forecasting, serving as an accelerator for the harder edges of healthcare, logistics, and environmental planning. Cloud-edge setups provide the flexible backbone, balancing global-scale analysis with local, split-second reactions that keep things grounded rather than abstract.

In the end, cybersecurity rounds out the picture, much like an immune system in any living organism. It quietly defends against the growing risks of constant connectivity, helping maintain stability even as threats evolve. These eight elements do not just stack up as a tech catalog; together they form a socio-technical base for life in what is billed as a Super-Smart Society. Each one contributes to efficiency, sustainability, fairness, and toughness in its own way. Still, it is their interplay that hints at the fuller vision, a human-centered setup where progress might actually lift dignity, inclusion, and the planet's long-term health, rather than leaving them as afterthoughts.

References

1. Cabinet Office, Government of Japan. (2016). *Society 5.0: A people-centric super smart society*. Government of Japan.
2. Fukuyama, M. (2018, July/August). Society 5.0: Aiming for a new human-centered society. *Japan SPOTLIGHT*. Japan Economic Foundation.
3. European Commission. (2021). *Industry 5.0: Towards a sustainable, human-centric and resilient European industry*. EU Publications.
4. Schwab, K. (2016). *The fourth industrial revolution*. World Economic Forum.

5. Xu, M., David, J. M., & Kim, S. H. (2018). The fourth industrial revolution: Opportunities and challenges. *International Journal of Financial Research, 9*(2), 90–95.
6. Lee, J., Bagheri, B., & Kao, H. A. (2015). A cyber-physical systems architecture for Industry 4.0-based manufacturing systems. *Manufacturing Letters, 3*, 18–23.
7. Arute, F., et al. (2019). Quantum supremacy using a programmable superconducting processor. *Nature, 574*(7779), 505–510.
8. Tapscott, D., & Tapscott, A. (2016). *Blockchain revolution: How the technology behind bitcoin is changing money, business, and the world.* Portfolio/Penguin.

Chapter 3
Innovations for a Greener Society

3.1 Introduction

A sustainable society is one in which human needs are met without compromising the planet's health and the well-being of future generations. In the context of "*A Journey Towards a Human-Centric World*," this chapter examines how sustainability can be embedded in communities, economies, and environments. It emphasizes futuristic, innovative, and responsible decision-making to create a world that prioritizes people while safeguarding natural systems. By addressing key areas such as energy, agriculture, transportation, governance, and social equity, this chapter offers a roadmap for building adaptable societies that flourish in harmony with the Earth.

The situation is critical and demands immediate attention. Climate change, resource depletion, and social inequalities pose significant challenges that require collective action. These challenges also present opportunities to reimagine how individuals live, work, and communicate. From renewable energy systems powering smart cities to sustainable agriculture ensuring food security, innovative solutions are already transforming communities. This chapter highlights practical strategies and illustrative examples that demonstrate how human-centric values, such as fairness, collaboration, and supervision, can drive sustainable progress. The synergy of pillars of sustainability is shown in Fig. 3.1.

Sustainability means creating a society that stays balanced. Environmental sustainability ensures the protection of ecosystems and the wise management of resources. Social sustainability promotes equity by ensuring that all people have access to opportunities and basic needs. Economic sustainability means growing the economy in ways that don't harm people or the environment, while still protecting resources for the future. It focuses on the efficient use of resources, the creation of fair employment opportunities, and the reduction of inequality, thereby benefiting everyone. In simple terms, it's about building wealth today without taking away opportunities for future generations. By integrating these pillars of sustainability,

S. Shukla et al., *Society 5.0*, SpringerBriefs in Applied Sciences and Technology, https://doi.org/10.1007/978-3-032-22164-3_3

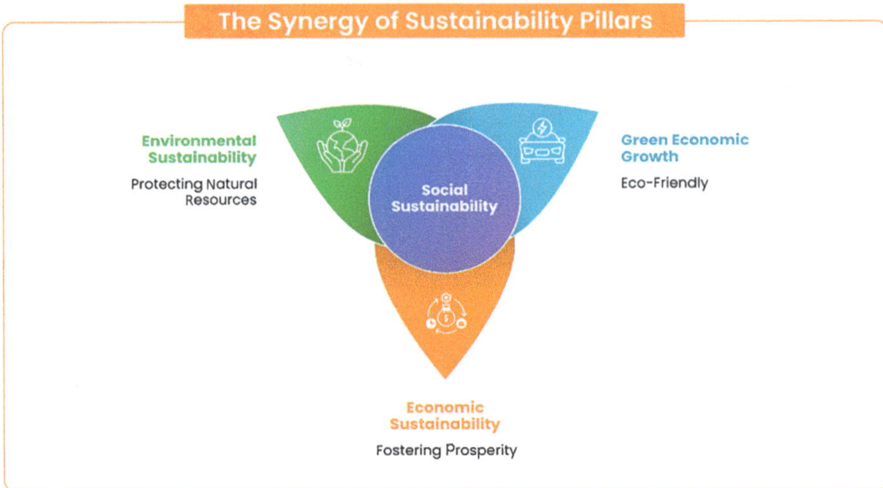

Fig. 3.1 Three pillars of sustainability for a Greener Society

one can create systems that support both human growth and planetary health. This chapter examines how green technologies, inclusive policies, and cultural shifts can work together to achieve this balance, offering hope for a future in which sustainability is a way of life.

The following sections explore critical components of a sustainable society, including green energies and smart cities, renewable energy systems, sustainable agriculture and water management, eco-friendly transportation, governance for sustainable development, and social equity. Each section underscores the need for collaborative, forward-thinking approaches to address global challenges. Together, these elements form a vision of a human-centric world where innovation and responsibility lead to a healthier, more equitable planet for generations to come.

3.2 Green Energy and Smart Cities

The transition toward a sustainable society relies primarily on integrating green energy systems and evolving smart cities. Green energy sources such as solar, wind, hydropower, and bioenergy aim to minimize carbon emissions to ensure long-term energy security. These renewable sources reduce dependence on fossil fuels, mitigate climate impacts, and promote circular economic models. As energy demand continues to rise, efficient energy conversion, storage, and grid management are essential for maintaining ecological balance and supporting the Sustainable Development Goals.

3.2.1 Introduction to Green Energies and Smart Cities

Smart cities serve as dynamic frameworks for implementing these clean energy solutions through advanced technological infrastructures. They encompass smart transportation, energy-efficient buildings, real-time data analytics, and digital governance mechanisms. By embedding sensors, Internet of Things (IoT) devices, and AI-driven management systems, smart cities optimize resource utilization, reduce waste, and enhance citizens' quality of life. The convergence of green energy technologies and smart city initiatives fosters resilient urban ecosystems that can adapt to environmental challenges while maintaining social inclusivity. Together, they constitute the foundation of a sustainable society in which technology, the environment, and human well-being align to create a future defined by efficiency, equity, and ecological harmony.

3.2.2 Role of Green Energies in Powering Smart Cities

Urban areas around the world are increasingly under pressure due to rapid population growth, climate change, and resource scarcity. As a result, green energy derived from low-emission sources such as wind, solar, hydro, biomass, and geothermal is essential to achieve urban sustainability. Nowadays, various systems use green energy to help reduce the carbon footprint. Similarly, smart cities have access to digital infrastructure such as the Internet of Things (IoT) and Artificial Intelligence (AI) to optimize urban services. The integration of green energy into urban systems clearly supports global sustainability goals, particularly SDG 7 (Affordable and Clean Energy) and SDG 11 (Sustainable Cities and Communities), encouraging cities to be both environmentally responsible and technologically advanced.

Green energy plays a significant role in powering the infrastructure and digital ecosystems of smart cities by providing reliable, decentralized power sources. Solar panels, wind turbines, and bioenergy systems supply renewable electricity that supports smart transportation networks, smart buildings, and automated urban services. These sustainable energy systems significantly reduce greenhouse gas emissions, helping cities transition toward low-carbon economies while improving air quality and public health.

In smart cities, the integration of green energy is enhanced by advanced technologies, such as smart grids and AI-based energy management systems. These tools balance energy demand and supply in real time, ensuring efficient distribution and minimal wastage. Smart meters and IoT-enabled networks enable residents and municipal authorities to monitor consumption patterns, thereby fostering energy-conscious behaviors. Additionally, microgrids powered by renewable sources strengthen energy resilience by enabling localized generation and storage, which is crucial during power disruptions. Through the integration of green energy and smart city innovation, urban areas become more sustainable, resilient, and self-sufficient. The result is a holistic ecosystem in which renewable energy not only powers daily operations but also drives technological advancements and promotes environmental supervision for future generations.

Green energy provides flexibility through rooftop solar, urban wind turbine installations, micro-hydro systems, and waste-based biomass systems, which complement the needs of a smart city's energy system. The installation of rooftop photovoltaic (PV) panels at homes, offices, and schools supports localized electricity generation that can be matched to real-time requirements and demand. Wind turbines, particularly those located on the outskirts or offshore, enhance the stability and consistency of solar power output. Continuous collaboration is required to develop a sustainable system, such as biomass plants that process organic waste and convert waste streams into power. Buildings equipped with smart, predictive sensors and control systems can help adjust lighting, cooling, and heating. When these renewable sources are at peak power (e.g., during high sunlight or strong winds), the energy is captured and used efficiently rather than wasted. Transportation options like electric buses, solar-powered EV buses, and renewable-powered metro networks help create low-carbon mobility while using electricity grids. The concept of decentralized energy systems increases the probability that neighborhoods can remain operational during grid outages.

3.2.3 Integration of Smart City Technologies and Green Energy

The integration of green energy into smart cities depends on advanced technologies, such as smart grids. The integration of smart electricity networks enables bidirectional energy flow and maintains a stable balance between demand and supply. Renewable energy can be stored in smart grids via on-site batteries or distributed energy resources, thereby reducing waste. The integration of AI and ML enables the analysis of these data streams, the forecasting of energy usage, the optimization of storage, and the monitoring of scheduled maintenance. Blockchain technology in renewable energy enables peer-to-peer energy trading, allowing businesses or households to sell surplus renewable energy directly to consumers. The architecturally feasible designs include energy-positive designs, green roofs, buildings integrated with photovoltaics [1], and high-efficiency materials that enable buildings to function as active hubs of energy rather than consumers. The combination of adaptive management and green generation helps create an urban environment that operates as a smart, energy-aware entity.

3.2.4 Challenges and Future Directions

Despite the integration of smart technologies and green energy, several implementation challenges remain. In developing cities, there are significant capital costs of renewable energy, smart grids, and IoT infrastructure. Storage limitations are a concern for urban renewable energy systems. There is no end-to-end support for peer-to-peer trading or decentralized energy generation under the regulatory frameworks. Public acceptance and technical capacity also vary drastically. The cybersecurity threat in smart systems is high and poses a significant risk if not addressed promptly.

However, upcoming innovations promise to enhance the reliability and affordability of community microgrids, AI-powered grid-based orchestration, and advanced battery technologies. One real-world example of this is Masdar City (UAE), which integrates solar power, electric vehicles, and energy-efficient transport to create a low-carbon city [1]. The city uses smart systems for lighting and waste management, as well as district energy networks, to reduce emissions and improve delivery services. Another example is Singapore's Punggol Digital District, which integrates solar-powered systems, green buildings, and efficient district cooling into a smart grid, helping shape the future of sustainable urban living.

3.3 Renewable Energy Systems

Renewable energy systems play a crucial role in making the world cleaner and greener. They replace fuels such as coal and oil with energy from natural sources, including sunlight, wind, water, plants, and geothermal heat. These sources are inexhaustible and can be used to generate green energy. Using renewable energy helps reduce air pollution and climate change, and keeps the energy supply safe and steady. With new technologies such as smart grids, energy storage batteries, and Internet of Things (IoT) devices, this energy can be used efficiently and without loss. Renewable energy also creates new jobs, supports both cities and villages, and helps people produce their own energy. As the world works to reduce carbon emissions, renewable energy is central to innovation, assisting individuals to protect the environment while improving technology and quality of life. Renewable energy systems play an essential role in this shift. These are some of the resources that are inexhaustible and abundantly available over human timescales. They promise a sustainable use of energy, unlike fossil fuels, as in the case of solar energy systems which convert sunlight into electricity using semiconductor materials via photovoltaic effect. Rotor-blade wind energy control systems capture the kinetic energy of moving air masses and convert it into mechanical energy, which is converted into electricity. Geothermal energy-powered systems extract heat from Earth's interior and use steam to generate power. Algae, forest waste, and crop residues are examples of organic matter components of biomass-based systems that are burned and gasified to produce usable energy. For electricity production, energy sources such as tidal power and wave energy are considered renewable energy systems because they harness the ocean's motion. Each system has a unique operating style that depends on the natural cycle and the physical forces available in the environment.

3.3.1 Global Energy Landscape

A fundamental transformation is underway in the global energy sector, driven by the immediate need to address climate change, improve air quality, and ensure long-term energy security. Fossil fuels such as coal, oil, and natural gas have powered industry and transportation and have become the primary energy source in recent years. However, these factors are affecting the environment. Global warming is accelerating, primarily due to rising greenhouse gas emissions. At the same time, water and air quality issues are causing serious health problems. There is a need to limit the rise in global temperature. The primary goal of the Paris Agreement, according to the Intergovernmental Panel on Climate Change (IPCC), is to limit the increase in global average temperature to 1.5 °C above pre-industrial levels. It plays a vital role in shaping how the world produces, distributes, and consumes energy.

3.3.2 Energy Storage and Grid Technologies

One significant challenge for renewable energy is the variability of its output. Wind power is entirely dependent on the weather, whereas solar power tends to drop at night because clouds block it. Without modern grid-based systems and sufficient storage space, there will be no continuous electricity supply. Whenever output declines, storage plays a significant role, helping to buffer excess renewable energy typically produced during periods of high production. Because of their low cost and ability to store large amounts of energy, lithium-ion batteries can now be considered a top option for both small and large power plant systems. Pumped hydro storage, widely used worldwide uses surplus power to pump water uphill and releases it later through turbines when demand is high. Thermal storage typically stores cold or heat for later use and is often paired with heating or cooling networks or power plants. Excess renewable energy is converted into fuel that later powers factories, transport systems [2], and electronic systems. It can be achieved using hydrogen produced by electrolysis, an emerging option for this process. One of the most demanding challenges today, in addition to storage, is the transition from legacy electrical grids to smart systems, which is considered vital for managing renewable energy. Smart grid systems employ advanced meters with automated digital controls that can integrate with local energy sources, thereby improving stability. They enhance system reliability and enable it to adapt to users' energy demand by creating microgrids that supply local power to communities during blackouts. Building stronger power lines facilitates the transmission of renewable energy over longer distances.

3.3.3 Environmental and Financial Aspects

Renewable energy systems offer various environmental advantages, including lower greenhouse gas emissions and reduced pollution, compared with traditional fossil fuels. Technologies like wind, solar, hydro, and geothermal energy are vital to promote global decarbonization and address climate change. However, they also raise environmental concerns, including habitat disturbance, land-use issues, risks to biodiversity, water-use demands, and occasional noise. To minimize these challenges, careful planning, control measures, and thorough environmental assessments are necessary to safeguard local ecosystems and communities.

The renewable energy system requires a significant initial investment in infrastructure and in grid integration. Successful investment in renewables depends on supportive public policies, financial incentives (like grants, tax benefits, and green tariffs), favorable market conditions, and easy access to financial support. In developing countries, the scale and speed of renewable energy system adoption depend on the overall implementation costs and investment risks. Renewable energy is becoming more affordable due to improved grid management, enhanced energy storage, and new market rules that enable renewables to compete more effectively with traditional fossil fuels. There are substantial economic and environmental benefits. For example, in Germany, the Energiewende policy is moving the country away from nuclear power and encouraging investments in energy efficiency, renewable energy, and modern grid systems. Costa Rica has also been producing almost all of its electricity from renewable sources, such as wind, hydropower, and geothermal energy, for many years. In India, large solar parks, such as the 750 MW Charanka Solar Park in Gujarat, are helping meet industrial power needs, create jobs, and expand electrification.

3.3.4 Innovation and Challenges

Innovations in renewable energy are quickly changing how the world generates and uses power. New developments include improved solar and wind panels and turbines, advanced battery storage systems such as lithium-ion and solid-state batteries, AI-enabled smart grids, and the growing use of green hydrogen as a fuel. Technologies such as perovskite solar cells, floating solar farms, and digital twin systems are enhancing the efficiency and scalability of renewable energy. Blockchain technology supports peer-to-peer energy trading, improving transparency and trust.

Despite these advances, several challenges remain [3]. Solar and wind power are intermittent, which makes it more difficult to maintain grid stability. High initial costs, the need for strong government policies, limited resource availability, and environmental concerns, such as land use and raw material sourcing, also pose barriers. Technical issues, such as battery recycling and hydrogen production, as well as economic and policy hurdles, require further research and development.

Examples of renewable energy systems include large offshore wind farms in Europe and Asia that supply substantial power to the grid. Solar parks paired with battery storage in India and the southwestern U.S. provide reliable energy even after sunset; green hydrogen pilot plants in countries like Germany and Australia aim to support carbon-free industries; and blockchain-based peer-to-peer energy trading models in Australia and Germany enable communities to share renewable electricity.

3.4 Sustainable Agriculture, Water Management, and Food Systems

Sustainable agriculture, frequently defined as Agriculture 5.0, focuses on improving farming by prioritizing people, nature, and long-term needs. It addresses two main global problems: rising food demand and a shortage of farm labor. Improving farming practices can produce sufficient food while protecting natural resources. One vital component of this change is improved water use. Farmers now rely on real-time information about soil and weather conditions to decide when and how much to irrigate. It helps avoid overwatering, saves water, and improves crop health. Such mindful use of resources reduces waste and decreases costs for farmers, mainly in regions facing water scarcity. Sustainable agriculture [4] also improves the flow of food from farms to consumers. Better coordination between farmers, storage facilities, transport systems, and markets reduces spoilage and delays. Transparent supply chains help ensure food quality and safety while allowing farmers to respond to changing consumer needs, such as demand for fresh, local, or customized produce.

Another pivotal goal is to make farming more adaptable. By reducing reliance on manual labour and on weather-related disruptions, farms can continue operating even during events such as climate or health crises. This stability is paramount as the global population continues to grow and food demand increases. Overall, this approach supports a sustainable future in which agriculture meets human needs without harming the environment. It combines economic growth with social well-being, ensuring that agricultural progress benefits farmers, consumers, and ecosystems.

3.4.1 Sustainable Agriculture Practices

Sustainable agriculture focuses on farming practices that protect the environment while still producing sufficient food for the present population without compromising resources for future generations. The main idea is to use natural methods that maintain healthy soil and protect biodiversity. One widely adopted sustainable practice is crop rotation, in which farmers grow different crops in the same field across seasons. It helps improve soil nutrients and disease cycles and strengthens the farm ecosystem. Farmers also diversify the fields by adding legumes, pulses, or cover crops. These crops naturally add nitrogen to the soil, reduce the need for chemical fertilizers, and help prevent soil erosion. As a result, the soil becomes richer, healthier, and more productive over time. Conservation tillage is another best practice that involves minimal soil disturbance. It maintains soil moisture, increases organic matter, and protects microorganisms that support plant growth. These methods work well across diverse climates and farm sizes, making sustainable agriculture practical in a wide range of agricultural settings. Sustainable farming also includes Integrated Pest Management (IPM), which employs organic inputs such as compost and biofertilizers rather than relying heavily on pesticides. It reduces environmental pollution and supports species such as pollinators, thereby promoting holistic farm health.

Another approach, agroforestry, combines trees with crops and livestock. Trees store carbon, enrich habitats, improve soil quality, and offer farmers additional income through fruits, fodder, and timber. Native knowledge and community-based practices are increasingly being adopted alongside modern technologies, such as precision farming, thereby increasing productivity while remaining environmentally responsible. For example, a farmer growing rice and vegetables may rotate crops annually and plant legumes during the off-season. They might use compost instead of chemical fertilizer and maintain patches of trees around their farm. Over time, their soil becomes more fertile, water use becomes efficient, and crop yields remain stable even during climate change. Sustainable agriculture, therefore, helps protect the environment, strengthens local communities, and ensures food security for future generations.

3.4.2 Sustainable Water Management Strategies

Sustainable water management strategies are essential for building resilient agricultural systems that balance productivity while preserving ecosystem health. These strategies aim to optimize the use of freshwater resources, reduce waste, and maintain the ecological functions that support farmers. By adopting precise irrigation technologies, landscape-level conservation practices, and data-driven planning, farmers can efficiently utilize water resources while safeguarding the environment. Core techniques in sustainable water management include drip irrigation, microsprinklers, and smart irrigation systems, which supply water directly to the plant root zone. It minimizes losses from evaporation and overflow, ensuring water usage only where needed. Rainwater harvesting and storage enhance water availability during drought days and reduce reliance on groundwater. Today, farms integrate soil moisture sensors, remote sensing data, and weather-based irrigation scheduling to make informed decisions on water allocation and quantity, resulting in improved yields and substantial water savings.

Landscape-based strategies further strengthen these efforts. Agroforestry systems, vegetative buffer strips, and riparian zones promote infiltration, control erosion, and enhance groundwater recharge. These measures also support biodiversity and stabilize soils. Sustainable water management also requires watershed protection, the use of drought-tolerant crop varieties, and organic farming approaches. The adoption of drought-resistant crop varieties and the protection of upstream reservoirs ensure that agricultural systems remain productive despite climate-related water challenges. Policy frameworks and community participation play a crucial role. Local water-sharing agreements, farmer cooperatives, and watershed governance initiatives encourage equitable water distribution and collective resource management. Emerging innovations, including wastewater recycling, the use of treated greywater, and robotic water-monitoring systems, are enabling farmers to conserve water while maintaining productivity. For example, in Karnataka's semi-arid regions, farmer groups have combined drip irrigation with joint rainwater harvesting structures and sensor-based irrigation scheduling. This integrated system has significantly reduced water consumption, stabilized crop yields, and improved groundwater levels, demonstrating how sustainable water management strategies can support both agricultural growth and environmental stability.

3.4.3 Sustainable Food Systems

Sustainable food systems consist of the processes, people, and resources involved in producing, processing, distributing, consuming, and managing food waste. Their goal is to ensure food security and good nutrition while protecting environmental, economic, and social systems for future generations. These systems connect key sectors, such as agriculture, water management, trade, energy, and public health. Sustainable food systems promote fairness, flexibility, and efficient use of natural resources from the farm to the consumer. Agroecological approaches form an essential foundation of sustainable food systems. Practices such as diversified farming, urban agriculture, local food networks, and short supply chains strengthen ecosystem services and support biodiversity. These models create circular flows of nutrients and energy by recycling organic matter and reducing dependence on long-distance transportation. As a result, they help lower greenhouse gas emissions, minimize food waste, and encourage cooperation and knowledge exchange among farmers, consumers, and communities.

Technological innovation also plays a subsequent role. Digital tools, supply-chain transparency, and improved food-labeling systems increase consumer trust by providing accurate information about the food. These tools help individuals make healthy, ethical, and sustainable food choices. Sustainable food systems also emphasize social well-being. They promote fair wages for farmers and encourage responsible consumption. It ensures that everyone has access to safe, nutritious, and culturally acceptable food. By combining efforts across education, health, business, and policy, these systems build strong resilience against disruptions such as extreme weather, supply chain breakdowns, and public health crises. Ultimately, they work to reduce environmental impacts while maintaining a stable and nutritious food supply. For example, a city that practices a sustainable food system promotes community gardens, supports local farmers' markets, and uses digital tracking for food safety. Residents have access to fresh, local produce; farmers receive fair income; and the city reduces food miles, energy use, and waste. This integrated approach strengthens both community health and environmental sustainability.

3.4.4 Challenges

Despite the presence of sustainable agriculture, integrated food systems, and efficient water management systems, significant challenges remain. Climate change is associated with unpredictable weather, higher temperatures, reduced rainfall, and increased flooding, disrupting agriculture and reducing global yields. These impacts reduce yields by up to 25% and damage 30% of fertile land [5], mainly in climate-vulnerable regions. Loss of pollinators and greenhouse gas emissions from farming worsen the problem. For example, unseasonal heavy rains in South Asia destroy rice seedlings, delaying planting and reducing harvests. According to the internet, approximately 31% of people in Africa lack access to basic drinking water. Since agriculture uses over 70% of global freshwater [6].

Climate-driven droughts and uneven rainfall increase the risk of crop failure. In arid regions of Africa, farmers rely on over-pumped borewells that run dry during drought, resulting in crop failures. There is a threat to the long-term availability of water, and soil degradation continues to reduce the overall productivity of farmland. There may also be problems arising from over-extraction of groundwater, threatening the future availability of water in that region.

Overuse of land, soil erosion, monocropping, and chemical fertilizers are degrading soils worldwide. By 2030, approximately one-third of topsoil may be severely degraded, leading to yield losses of up to 50% in some areas and increased reliance on synthetic inputs. As in continuous maize farming in Latin America, it depletes soil nutrients, forcing farmers to use more fertilizer each year. Small-scale farmers continue to face poverty, limited market access, high input costs, and inadequate financial support. Gender and land tenure inequality further weaken rural resilience. For example, women farmers in many regions cannot access loans for seeds or equipment because they don't hold land titles. Innovations such as precision agriculture, climate-smart seeds, and digital advisory tools are rapidly growing, but adoption remains uneven due to high costs, limited training, and weak rural infrastructure. For example, farmers in remote villages cannot use weather-forecasting apps due to poor network connectivity. Fragmented policies, misaligned subsidies, and limited support for regenerative or agroecological practices slow down sustainable transitions. In the future, short-term policy cycles and weak farmer participation may hinder the development of coordinated, long-term solutions. Subsidies that favor chemical fertilizers over organic alternatives discourage farmers from shifting to regenerative farming.

3.5 Green Transportation

Transportation is a key aspect of building a greener society. It focuses on moving people and goods in ways that reduce pollution, save energy, and protect health. Traditional transport, powered by petrol and diesel, emits substantial greenhouse gases and harmful air pollutants. It also creates noise, traffic congestion, and unsafe streets in many cities. Sustainable transport offers a better path. It includes walking, cycling, shared public transport, and greener vehicles such as electric cars and buses. These options can cut emissions, improve air quality, and make communities more inclusive and livable. This section examines why change is necessary, the existing solutions, and how policy, technology, and personal choices can shape the future of mobility.

3.5.1 Greener Mobility for People and Planet

Traditional transportation relies heavily on fossil fuels, such as petrol and diesel. This reliance produces significant greenhouse gas emissions, particularly carbon dioxide (CO_2), which drives climate change. Globally, the transportation sector accounts for nearly 25% of CO_2 emissions. Additionally, vehicles emit harmful air pollutants, including nitrogen oxides and other fine particles. These pollutants cause respiratory illnesses and heart disease. Vehicle noise adds to urban stress and disrupts communities. Traditional car-centric city designs separate homes, workplaces, and shops. It makes daily commuting difficult and limits mobility for non-drivers, including low-income individuals and the elderly. Climate change worsens these issues, as extreme weather disrupts transport networks. Recognizing these problems, cities such as London have implemented Ultra-Low Emission Zones to reduce pollution in dense areas. Bogotá hosts weekly Ciclovia events in which streets are closed to motor vehicle traffic on Sundays and public holidays to encourage cycling and walking, thereby promoting health and cleaner air. These examples show the urgent need for sustainable, inclusive mobility that cuts emissions and improves public health and accessibility.

3.5.2 Sustainable Transport Solutions

Sustainable transport includes walking, cycling, public transit, and electric vehicles. A mix of sustainable transport options forms the backbone of eco-friendly mobility. Active transportation, such as walking and cycling, requires no fuel and emits no pollution. It also delivers health benefits through exercise and reduced stress. Copenhagen has invested heavily in bicycle infrastructure, resulting in nearly 50% of its residents commuting by bicycle daily. Public transit, including buses, trams, and subways, moves large numbers of people efficiently and produces much lower per-passenger emissions than private cars. Curitiba, Brazil, is renowned for its bus rapid transit system, which serves as a prototype for cities worldwide. Electric vehicles (EVs) are gaining prominence, powered by electricity from renewable sources. Shenzhen, China [7], operates one of the world's largest fleets of electric buses, drastically cutting urban emissions. While EVs reduce local air pollution and noise, they should complement, not replace, walking, cycling, and public transit to form an integrated system that serves diverse needs sustainably. Together, these options create a diverse toolkit to cut the environmental footprint of transportation.

3.5.3 Policies and Designs for Green Transport

Infrastructure and policy create the foundation for sustainable transport options to flourish. Compact mixed-use urban designs bring homes, workplaces, schools, and shops closer together, enabling shorter trips that encourage walking, cycling, and public transport use. Portland, Oregon, is a prime example of transit-oriented development aligning housing with frequent transit service. Streets designed as complete streets allocate safe space equally to pedestrians, cyclists, and vehicles, improving safety and comfort. Governments are bringing policies to support green transport. Emission standards, incentives for EV ownership, subsidies for transit, and limits on polluting vehicles create financial and regulatory motivation to choose cleaner options. Investments in public transport expansion, cycling infrastructure, and EV charging also enhance the appeal and reliability of sustainable transport. Digital innovations like smart traffic lights and real-time transit tracking applications enhance the user experience. These combined efforts improve sustainable transport choices for the planet and make them easier for the public.

3.5.4 The Future of Green Transport

The future of transportation is promising, with numerous innovations. Next-generation batteries increase EV range and reduce charging time. Hydrogen fuel cells could power trucks and buses over longer distances where batteries fall short. Autonomous vehicles may reduce accidents and improve traffic flow. Upcoming urban air mobility concepts, such as flying taxis, aim to minimize ground-level congestion. Digital platforms integrate taxi-booking, cycling, and transit into a seamless experience known as Mobility as a Service (MaaS). However, technology alone will not solve all challenges. Individual and collective actions matter. People choosing to walk, bike, carpool, or take transit help reduce emissions and traffic. Employers who promote flexible working hours and telecommuting reduce peak commute loads. Communities advocating for bike lanes, park-and-ride facilities, and green-vehicle incentives create demand for improved services. Sustainable transportation is a shared endeavor that integrates innovation, policy, and everyday choices to achieve healthier, cleaner, and livable cities. For example, Shenzhen, a large Chinese city, became the first city in the world to replace all 16,000 of its public buses and 22,000 cabs with electric vehicles.

The city has installed thousands of charging stations and deployed technology to track battery levels. Carbon emissions consequently decreased by over 440,000 tons annually. It demonstrates that when a community has appropriate planning, government assistance, and an emphasis on cleaner air, green transportation can succeed.

3.6 Governance and Policy for Sustainable Development

Governance and policy are the backbone of any action toward sustainable development. More than just creating rules, strong governance builds systems that promote transparency, fairness, and accountability across sectors and levels of government. National policies set the overall direction and align with global sustainability goals, but real change happens when state, city, and community-level institutions adapt these plans to local requirements. International cooperation also plays a key role. Countries collaborate through global agreements supported by organizations that offer funding, technical assistance, and shared expertise. Still, several challenges persist, such as coordination gaps, limited financial resources, and shifting political priorities, which can slow progress and dilute long-term efforts. In the future, the path to sustainability requires governance that is adaptable, inclusive, and open to innovation. Building stronger partnerships between global decision-makers and local communities will be essential for addressing complex environmental, social, and economic problems. Overall, effective governance and well-designed policies are crucial tools for addressing current sustainability challenges and creating an adaptable, equitable future for everyone.

3.6.1 Foundations of Sustainable Governance

Sustainable governance is a system in which environmental, social, and economic goals are integrated into everyday decision-making across public and private institutions. It not only represents rules and regulations but also a culture of transparency, accountability, and inclusiveness. A sustainable governance framework seeks to use resources wisely over the long term while balancing social equity and the environment. Achieving this requires collaboration among stakeholders, including governments, industries, civil society, and local communities, as sustainability challenges rarely exist within a single sector. Issues related to energy, water, health, agriculture, and urban development are deeply interconnected. This overlap often creates complexity; to address it, governance systems are needed that are flexible, adaptive, and responsive to emerging risks and shifting conditions.

Sustainable governance provides the structure and guidance needed for societies to meet today's demands without compromising the well-being of future generations. It embeds sustainability into leadership, planning, and management, making long-term adaptability a shared commitment rather than an afterthought.

3.6.2 Central and State Level Initiatives

Sustainable development strategies are most effective when implemented across multiple levels of governance. National governments typically develop comprehensive policies that align with climate objectives, energy roadmaps, legislation, and global goals. In line with the SDGs, most countries have developed climate action plans, renewable energy policies, and social inclusion frameworks to address core sustainability challenges. These action plans often look impressive on paper. However, in practice, responsibilities shift to states, districts, and city administrations that interact directly with communities. Local governance enables implementation tailored to local geography and circumstances, engages citizens, customizes programs, and manages resources efficiently. For example, cities like Curitiba, Brazil, are world-renowned for pioneering integrated urban sustainability practices. It did not become a model for sustainable urban planning overnight; it arose from a series of practical decisions, such as bus corridors, green spaces, and community involvement, that slowly added up to become a sustainable city. In a hierarchical governance approach, coordination is essential to prevent duplication of work and to ensure that all parties move toward common goals. In this model, progress may sometimes stall as everyone assumes someone else will handle the task.

3.6.3 International Cooperation for a Greener World

Global challenges such as climate change and biodiversity loss do not stop at national borders; therefore, countries need to work together. Agreements such as the Paris Climate Accord, the Convention on Biological Diversity, and the Sustainable Development Goals provide nations with a shared direction and a common framework for action. They outline shared commitments, measurement goals, and reporting obligations, thereby creating accountability for individual countries at the global level. International institutions such as the United Nations, World Bank, and Global Environment Facility support this implementation through funding, technical expertise, and monitoring. They also help countries, particularly developing ones, build the skills and systems needed to meet sustainability targets. These institutions foster partnerships among governments, civil society, and the private sector. Even with strong frameworks and resources in place, success still relies on political commitment, fair participation, and effective enforcement. Bridging the gap between international goals and local realities remains a persistent challenge. For example, global collaboration, primarily through the C40 Cities Climate Leadership Group [8], drives economic growth in cities like Shenzhen.

To achieve the Sustainable Development Goals (SDGs), this international group of around 100 mayors collaborates. Cities such as London, Paris, and Los Angeles exchange procurement strategies, technological data, and policy sandboxes to advance zero-emission transportation under the C40 Fossil Fuel-Free Streets Declaration. By enabling cities to learn from the infrastructure issues of early adopters, this worldwide plan avoids "reinventing the wheel." In Society 5.0, C40 ensures that green transportation is a unified worldwide movement toward environmental sustainability rather than merely a local success of a shared strategy.

3.6.4 Challenges and Future Path

While governance systems and policies have improved over time, many implementation gaps persist, including limited coordination, resource constraints, and political fragmentation, which often impede progress. Social inequalities and competing economic interests can weaken the sustainability agenda. Data gaps and inadequate monitoring obstruct evidence-based adjustments. Building trust with communities and involving marginalized groups remains a critical challenge. Going ahead demands reimagined, more inclusive governance that is transparent, participatory, and adaptive to change. Integrating innovative tools, such as online platforms for public feedback, improved environmental monitoring systems, and integrated planning tools that connect sectors, can help. Strengthening linkages between local and global stakeholders, fostering closer cooperation among public, private, and civil society, and embedding sustainability within core institutions are essential steps. The future of sustainable governance lies not just in policies but in nurturing ethical leadership, shared responsibility, and a vision that balances People, Planet, and Prosperity (PPP). For example, in New York City's High Line project, the authority effectively accomplished SDG 11 (Sustainable Cities) by converting outdated infrastructure into green space. However, this also led to "green gentrification."

The absence of an integrated policy led to governance failure: SDG 10 (Reduced Inequalities) was violated because the project's intended low-income residents were displaced by the park's rapid rise in property values. It demonstrates that Society 5.0 governance necessitates a holistic policy design that upholds social equity while promoting environmental sustainability.

3.7 Social Equity and Sustainable Society

Society 5.0 prioritizes human needs such as happiness, fairness, and wellness over financial growth. It seeks to build a smart society in which physical and digital systems collaborate to enhance everyone's quality of life regularly, irrespective of age, gender, or geography. People participate in decision-making in this framework, and policies are designed to promote equity, transparency, and personal well-being rather than just business interests.

Reducing inequality, particularly the digital divide affecting underprivileged and rural populations, is the primary objective of Society 5.0. Due to limited internet connection and a lack of digital skills [9], many people fall behind. To solve this problem, approaches that promote equal opportunity for all individuals have been prioritized over systems that benefit specific individuals.

Additionally, this vision aligns with international objectives of the Sustainable Development Goals. Society 5.0 fosters innovations to address problems such as an aging population and environmental harm by promoting collaboration among governments, businesses, and communities. Overall, it promotes a robust, adaptable, and sustainable future in which technology benefits society as a whole [10].

3.7.1 Understanding Social Equity

Social equity refers to the fairness with which resources, rights, and opportunities are distributed within society. It recognizes that people have different starting points in life and may face obstacles due to their race, gender, income, or background. Simply treating everyone the same doesn't always work because not everyone has equitable access to basic needs such as education, clean water, or healthcare. In a sustainable society, social equity involves removing these barriers so everyone can thrive. It is not merely about fairness; it is about creating stronger communities in which people support one another and feel a sense of belonging. For example, some cities work to ensure that new parks and green spaces are built in areas with historically limited access, thereby improving the quality of life for all residents.

One practical example of social equity is Medellín's Metrocable System, which treated mobility as a social right by connecting remote mountainous regions to the city's business hub while minimizing spatial inequality. It is a strong example of how Society 5.0 defines social equity as a design problem in which technology and infrastructure are purposefully used to integrate marginalized populations into the core of society. A comparable example is Seoul's Digital Inclusion Policy. To provide economically disadvantaged individuals and senior citizens with equitable access to e-governance and public services, Seoul implemented a policy that included free public Wi-Fi zones, smart community centers, and mandatory digital literacy training. Technology serves as public infrastructure rather than a commercial product, illustrating how Society 5.0 implements equity through integrated digital engagement and physical connectivity.

3.7.2 Participation and Justice

A truly sustainable society ensures that everyone has a say in decisions that affect their lives. It means listening to the voices of neglected groups and including them in planning and policymaking. Social justice is part of this, ensuring systems are accountable and citizens' rights are protected. Local communities demanding better schools or cleaner streets exemplify participatory action. Such involvement helps build trust between citizens and governments, making sustainability efforts more effective. Justice also means addressing historic inequalities and restoring balance by supporting those who have been left behind. When social equity is prioritized, it promotes both fairness and stronger social connections.

Society 5.0 defines participation and justice as governance capacities in which digital systems mediate public interests before policy decisions are taken fairly.

According to Society 5.0, involvement and justice are governance capabilities that mediate public interests fairly through digital platforms before final policy decisions. In this, critical technology concerns are acknowledged at an early stage, digital technologies are used to gather citizens' opinions, and agreement is reached rather than bureaucratic domination. Taiwan's vTaiwan platform [11], which uses the AI-based digital platform Pol.is, system to cluster citizens' ideas and highlight areas of mutual agreement across opposing parties during debates on issues such as ride-hailing, legislation, and financial technology. Procedural fairness is vital because policy decisions are made through transparent public engagement, thereby limiting business dominance.

3.7.3 Overcoming Inequality

Inequality remains one of the biggest barriers to sustainability. Often, underprivileged groups face greater pollution and health risks, as well as fewer economic opportunities. To build a fair society, targeted policies are needed. These include fair wages, affordable housing, improved public transportation, and access to legal protections. Some countries have developed "just transition" plans to help workers move from polluting industries to green jobs without losing their livelihoods. A "just transition" plan is a strategy to help workers and communities shift smoothly from sectors that are heavily polluting, such as coal mining or oil production, to cleaner, greener jobs without losing their income or livelihoods. It involves retraining workers for new roles in renewable energy and other sustainable industries, providing financial support during the transition, and investing in local economies to create new opportunities. For instance, coal miners may receive training to become solar panel installers, with financial help to ease the transition. The goal is to ensure that, as the economy becomes greener, no one is left behind or suffers unfairly. Such plans balance environmental goals with social fairness, making the shift to a sustainable economy inclusive. This approach balances economic growth with social care. Achieving this balance is complex and requires sustained effort, but a society must be truly sustainable and inclusive.

One example of overcoming inequality is the DigiGaon initiative in India. Society 5.0, which addresses inequality by utilizing digital systems as social infrastructure to alleviate inequalities brought by connectivity and location. India's DigiGaon (Digital Village) initiative aims to reach remote communities by offering high-speed internet and common service centers as integrated service hubs. Residents of villages such as Dhanauri can access government services, digital banking, and telemedicine without traveling far, thereby reducing their reliance on cities. The elimination of geography-based inequality, in which access to knowledge, healthcare, and finance is no longer dependent on one's physical proximity to urban centers. It provides women and small farmers with digital literacy initiatives that enable them to participate directly in online marketplaces and precision agriculture platforms. This illustration demonstrates how Society 5.0 implements justice by integrating technology into daily activities to ensure equitable participation in economic and social systems.

3.7.4 Building a Stronger and Inclusive Future

Social equity plays a vital role in building strong communities that can withstand challenges such as climate change and pandemics. When everyone, regardless of their background or status, has fair access to opportunities and resources, communities become stronger and better prepared for crises. For instance, during the COVID-19 pandemic, neighborhoods with equal access to healthcare and social services outperformed those with long-term inequalities. Equitable societies ensure that everyone receives the help they need, access to learning opportunities, and community support to cope and bounce back. These balanced communities promote trust and cooperation, turning problems into opportunities to bring people together rather than widening the gaps.

New digital tools for community-led monitoring, such as citizens tracking water quality or air pollution, hold authorities accountable while empowering residents. Together, these tools build stronger connections between governments and communities. A sustainable society emerges from shared values such as kindness, respect, and collective responsibility. It protects both people and the environment, prioritizing long-term well-being. Ultimately, social equity ensures that progress benefits everyone, creating a greener, more equitable world. For example, Japan's Silver Tech technology is a real-world example of building a stronger and more inclusive future by transforming an aging population into an opportunity for innovation. With nearly one-third of the population aged 65 or older, the so-called "2025 Problem" raises concerns about labor shortages, a slowing economy, and rising healthcare costs. Japan adopts inclusive, holistic technologies to maintain engagement and respect for senior citizens rather than excluding them. In industry and healthcare settings, older employees and caregivers can safely perform physically taxing jobs by using assistive devices such as Cyberdyne's Hybrid Assistive Limb. Automated last-mile transport enables senior citizens with mobility limitations to reach remote areas, reducing loneliness and ensuring access to markets and services. Additionally, digital platforms transfer skills to newer employees and intelligent systems by preserving the implicit knowledge of aging expert artisans. Taken as a whole, these actions demonstrate how technology may actively support humanity in society by promoting a fulfilling work-life balance, enhancing efficiency, advancing social justice, and transforming aging from dependency to empowerment.

3.8 Sustainability Challenges in Growth-Centric Models

The shift exposes a fundamental conflict in conventional growth-centric models for a Super Smart Society: the dilemma of pursuing limitless economic growth within a finite ecological and demographic framework. The Society 5.0 mission for equity and environmental balance remains at odds with traditional industrial mindsets, which often associate success with expanding consumption and continuous production. The need for GDP growth has historically led to the marginalization of nonproductive populations and excessive resource exploitation in many aging countries, resulting in systemic vulnerabilities that technology alone cannot resolve. The "Super Smart" paradigm must separate economic growth from resource depletion to attain true sustainability, shifting the emphasis from quantitative growth to qualitative well-being. To do this, value chains must be significantly redesigned, shifting from planned obsolescence to reuse and recycling, and ensuring that the Cyber-Physical System's [12] energy-intensive computing demands do not jeopardize the very climate stability they are meant to preserve.

3.8.1 Limits of Expansion

Growth-centric models push for larger economies over the years, measuring success by rising GDP. Natural resources such as forests, water, and oil are used faster than they can be replenished. Factories produce products, but waste accumulates in landfills or oceans, contaminating soil and wildlife. Air is filled with smog from cars and factories, leading to health issues such as asthma in children. These models assume endless expansion, yet Earth has finite supplies. Growth-mindset people often ignore this, prioritizing quarterly profits over planetary health. Inequality rises; top earners grab most of the gains, while wages decline for others. During recessions, vulnerable people are laid off first. Sustainability urges rethinking: why chase infinite growth on a finite planet? When growth compresses nature, everyone loses; future generations inherit unproductive land. Innovative models measure well-being, not just wallets. Costa Rica demonstrates that prioritizing eco-parks over heavy industry, attracting tourists, and keeping forests intact. Growth without guardrails speeds toward collapse. The global semiconductor industry, particularly in Taiwan, exemplifies these limitations. The manufacturing of microchips, the foundation of the CPS, has reached a physical ceiling as the world desires an exponential rise in AI and "super smart" technology. Large volumes of pristine water and electricity are required to manufacture these chips. Taiwan had to make a crucial decision during the 2021 drought: safeguard local farmers' and residents' water rights or shift water to the rapidly expanding tech sector to sustain global supply chains. This "resource collision" demonstrates that Society 5.0's infrastructure cannot grow endlessly without compromising people's fundamental ecological needs. It highlights the "Growth-Centric Paradox," which calls for a shift toward sustainable, resource-neutral innovation, as the same technology intended to address societal problems could exacerbate resource scarcity if it continues to operate under a model of endless expansion.

3.8.2 Costs of Rapid Growth

Many growth models promise opportunity for everyone, but reality often unfolds a different story. Factories close in established areas as production shifts to cheaper labor elsewhere. As clean energy expands, coal workers lose jobs without proper support. The gap between the rich and the poor grows, and executives earn huge salaries while regular employees barely manage to pay their bills. Communities near factories face polluted water and air, while wealthier neighborhoods stay protected. In a few places around the world, rapid industrial growth has sometimes meant child labor and contaminated groundwater. Fair transition programs can help, such as training coal miners for jobs in solar energy, so incomes remain stable. Without such planning, frustration builds and takes the shape of revolt, as seen in the "yellow vest" protests in France over rising fuel costs. When systems prioritize profit, communities disintegrate, and people compete rather than support one another.

Women and marginalized groups are often affected the most, ending up in low-wage jobs as automation grows. Data shows that equitable places innovate more, with diverse teams solving problems creatively. Equitable systems invest in training, equitable wages, and the creation of green space. The problem persists, as leadership decisions often prioritize stakeholders over workers. Germany's cooperatives blend worker ownership with green technology and fairly share benefits. When economic growth ignores people, society suffers. Inclusive approaches restore divisions and shift the mindset from competition to shared growth.

3.8.3 Immediate Benefits and Future Liabilities

Leaders chase rapid profits, ignoring future risks. Fossil fuels look cheap today, but they carry massive costs later, including climate change, air pollution, floods, droughts, and rising sea levels. Growth plans appear attractive and economical today, but they erode our children's future. Rising sea levels are drowning islands and displacing millions of people. Warming oceans kill coral reefs, starving coastal fishers. Models assume tech fixes all, but batteries require rare metals, sparking new mining wars. People in oil-producing regions fear job losses and therefore delay adopting renewable technologies. Bhopal's gas tragedy ruined lives, proving that profit prioritized above safety often results in costly, irreversible consequences. Green bonds are used to channel economic support, such as loans, to fund projects that support the environment, including green energy, wastewater treatment plants, solar energy, eco-transport, and sustainable farming. When leaders chase numbers, the quality of life is compromised. A balanced model fosters creativity; Amsterdam's bike lanes promote cycling over car use, improve health, and reduce medical costs. Long-term planning turns a threat to shared heritage, such as Waste-to-Energy (WtE) plants. Sustainability pushes us to slow down and think wisely rather than rushing.

3.8.4 *Resistance to Policy Change*

Some dominant companies and influential individuals often resist policies necessary for sustainability, such as environmental regulations or fair taxation. They fear these rules will reduce profits or weaken their influence. For example, coal industry leaders try to convince civic leaders to protect coal jobs and delay the transition to sustainability. Because of this resistance, harmful practices continue. Factories still release waste because penalties are too low to stop them. In Brazil, some farmers resist forest protection to expand soy production, accelerating the destruction of the Amazon rainforest. In Europe, truck owners protest fuel taxes, saying the cost is too high. This resistance slows progress, keeps the air polluted, and deepens inequality. In this, wealthy groups benefit while poorer communities suffer the consequences.

To overcome this, everyone's voice needs to be heard, not only those with money and power. Leaders must listen to workers, local communities, and environmental advocates. Decisive government action can create fair rules that protect nature and support green jobs. Public involvement also matters. When people share stories about polluted air or sick children, change becomes real and urgent.

3.9 Conclusion

In the context of Society 5.0, the shift toward a "Greener Society" signifies an underlying shift toward a shared economic mindset rather than merely technological enhancement. The integration of Cyber-Physical Systems (CPSs) provides unprecedented tools for resource optimization across precision agriculture and smart grids, as discussed in this chapter. These advances nevertheless fall into a "Growth-Centric Paradox." The energy requirements of global AI infrastructure and resource conflicts in the semiconductor sector illustrate how technology intended to address social and environmental issues may inadvertently exacerbate resource shortages if it remains tied to the concept of limitless growth. The correct course of action requires a shift away from efficiency and toward "green innovation." To do this, circular value chains and resource-neutral technologies must be prioritized to separate economic growth from environmental degradation. The ability to create a "Super-Smart Society" that respects global limits will be the accurate measure of success in Society 5.0, not just the speed of technological throughput. We can ensure that Society 5.0 is both innovative and inherently sustainable by establishing a balance between high-resolution data and ethical responsibility, safeguarding the environment as the cornerstone of human well-being.

References

1. Faes, A., Virtuani, A., Quest, H., et al. (2025). Building-integrated photovoltaics. *Nature Reviews Clean Technology, 1*, 333–350. https://doi.org/10.1038/s44359-025-00059-9
2. SG Analytics Knowledge Team. (2025, January). *The rise of eco-friendly transport: From EVs to hyperloop*. SG Analytics. Retrieved from https://www.sganalytics.com/blog/eco-friendly-transport-evs-to-hyperloop/
3. Sobey School of Business. (2024). Chapter 1: Social challenges. In *Exploring sustainability*. Atlantic OER/Pressbooks. Retrieved from https://pressbooks.atlanticoer-relatlantique.ca/exploringsustainability/chapter/chapter-1-2/
4. Hiywotu, A. M. (2025). Advancing sustainable agriculture for goal 2: zero hunger—A comprehensive overview of practices, policies, and technologies. *Agroecology and Sustainable Food Systems, 49*(7), 1027–1055. https://doi.org/10.1080/21683565.2025.2451344
5. Zhao, C., Liu, B., Pautsch, G. R., Sappington, T. W., & Yang, S. (2021). *Without adaptation to climate change, 25% of global crop yields could be lost*. University of Venice Ca' Foscari News. Retrieved from https://www.unive.it/pag/16584/?tx_news_pi1%5Bnews%5D=10718&cHash=2a8f59af27d6825acd4b7f8785964a95
6. United Nations Educational, Scientific, and Cultural Organization. (2024). *Water demand and use*. UNESCO World Water Development Report. Retrieved from https://www.unesco.org/reports/wwdr/en/2024/s
7. World Bank. (2021). *Electric mobility: The case of Shenzhen*. World Bank Group. Retrieved from https://openknowledge.worldbank.org/handle/10986/35134 (Note: This is the official report detailing the conversion of 16,000 buses and 22,000 taxis.)
8. C40 Cities. (2021). *C40 Fossil Fuel-Free Streets Declaration: Transforming cities through zero-emission mobility*. C40 Knowledge Hub. Retrieved from https://www.c40knowledgehub.org/s/article/Fossil-Fuel-Free-Streets-Declaration
9. Majumder, S., & Dey, N. (2023). Manpower development as an element of sustainable development. In *The vogue of managing people in workplace* (pp. 103–129). Springer Nature Singapore.
10. Komine, T. (2015). *Thinking about the 2025 problem—Part 1: Population change and the 2025 problem*. International Institute for Population Sciences, IIPS Quarterly.
11. Hsiao, Y., Lin, S. Y., Tang, A., Narayanan, D., & Sarahe, C. (2018). *vTaiwan: Public participation methods on the cyber-physical frontier*. vTaiwan.
12. Patil, R. V., Ambritta, N. P., Mahalle, P. N., & Dey, N. (2022). Medical cyber–physical systems in Society 5.0: Are we ready? *IEEE Transactions on Technology and Society, 3*(3), 189–198.

Chapter 4
Applications of the Society 5.0 Framework

4.1 Education 5.0: Lifelong Learning and Skill Development

Education 5.0 represents a necessary transformation of the learning paradigm aimed at preparing individuals for the human-centered super-smart society. Disruptive changes, accelerated by events such as the COVID-19 pandemic, drive this evolution, merging technological innovations across cyber, physical, and biological domains. The fundamental goal is to use technology to address social problems and drive economic advancement. Precisely, Education 5.0 demands a shift from focusing only on technology integration to implementing holistic, effective, and outcome-driven strategies in higher education.

4.1.1 Strategic Shift and Holistic Transformation

The rapid advancements of the Fourth Industrial Revolution generate a risk of "social pain" due to the decoupling between new technological realities and outdated educational paradigms. Education 5.0 seeks to break this harmful cycle by using digital tools to re-couple learning with technological advancements. This approach mandates that Higher Education Institutions (HEIs) [1] swiftly adapt and transform their processes. Technological, human, organizational, and pedagogical drivers must holistically guide effective Digital Transformation (DT). Successful transitions, such as migrating to Agriculture 5.0, integrate cutting-edge technologies (AI/IoT) [2] with human welfare goals, ensuring sustainability through robust organizational policies and ongoing training. It ensures that the institution's response moves beyond external changes towards aligning educational practices with the dynamic demands of modern society.

S. Shukla et al., *Society 5.0*, SpringerBriefs in Applied Sciences and Technology, https://doi.org/10.1007/978-3-032-22164-3_4

4.1.2 Cultivating Twenty-First-Century Skills

A primary objective of Education 5.0 is to equip students with a comprehensive set of capabilities necessary for the future workforce, requiring continuous learning, unlearning, and relearning. Students must acquire cognitive, social, interpersonal, and technical skills to fulfil the new job demands. Key soft skills include critical and analytical thinking, communication, creativity, and problem-solving, all of which are needed to navigate a multifaceted, interconnected world. Technical skills, such as digital technologies, design, and computational thinking, are also essential. Additionally, nurturing the learning ability is strategic for the survival of HEIs in the current learning space. Promoting adaptability, stress tolerance, and flexibility is key to success in the future learning space. HEI must plan a faculty and alumni reskilling program through microcredentials or nanodegrees. It will be beneficial in terms of time, money, and effort, as they can complete these courses while working.

In Society 5.0, it is necessary not only to learn the basics of science and computing but also to engage in active communication, critical thinking, practice empathy, and understand a complex world. It empowers people to use technology confidently and contribute to effective decision-making.

Society 5.0 revolutionizes education by putting people before technology. It uses AI, VR, and data tools to build valuable skills for everyone. Education in Society 5.0 cultivates skills through logical reasoning, critical data analysis, and privacy awareness. It imparts cognitive strengths like creativity, adaptability, and collaboration. Learning becomes personalized through innovative platforms and simulations. Microcredentials promote lifelong learning and bridge digital divides to achieve equitable access.

4.1.3 Innovative Pedagogy and Agile Learning

The goal of Education 5.0 is to transform teaching by promoting innovative pedagogy rather than relying on traditional methods. Fundamental methodologies include personalized and adaptive learning approaches, immersive and experiential learning, inclusivity and accessibility, digital literacy, and Skills for Social Transformation (critical thinking, problem-solving, creativity, and lifelong learning). These methods promote learning by doing, helping people gain and use skills to tackle specific challenges such as climate change, agricultural modernization, and policy reforms. To manage complexity and ensure continuous improvement, the learning process should incorporate the agile paradigm through time-boxed sprints. It allows learning through short, continuous deliveries that reduce risks and enable the creative use of knowledge. Accordingly, methodologies such as portfolio-based grading, peer assessment, and real-world project outcomes will serve as criteria for grading and evaluation. Additionally, the framework should integrate humanistic and societal impacts directly into the core of technical education, such as STEAM (integrating the arts into STEM).

4.1.4 Leadership and Institutional Adaptation

Successful transition to Education 5.0 depends on responsive, active leaders who can proactively address challenges and opportunities. The key competencies required of HEI leaders to inculcate in the teaching and learning process are adaptability, strategic planning, digital capability, and social responsibility. These leaders must guide the institution with clear organizational strategies that align with individual employees' skills and promote continuous learning and innovation. Leaders should use resources wisely, support long-term innovation, and involve citizens in transparent validation. This proactive leadership is essential for driving the cultural and operational changes needed to advance toward Education 5.0. HEI leadership mandates investment in professional development to train faculty in adaptive pedagogies and new technologies, like AI and Cyber-Physical Systems. This strategic adaptation transforms the curriculum to focus on essential human skills, such as critical thinking, ethical reasoning, and integrating sustainability across all subjects. This holistic shift ensures students are fully equipped to lead ethically and thrive in the human-centered Society 5.0.

4.2 Smart Cities: Urban Planning for Sustainable Living

The growth rate of urbanization necessitates sustainable urban development. It positions smart cities as an essential framework for addressing interconnected environmental, economical, and social challenges. Smart city [3] initiatives utilize advanced technology enhance the quality of life and sustainability, guided by the global values of Society 5.0.

4.2.1 The Foundational Role of Society 5.0 in Smart Cities

The development of smart cities must operate within frameworks that balance multiple dimensions of progress, not only the technological aspect. It should work toward a comprehensive vision of societal benefit. Society 5.0 is characterized as a human-centric ecosystem that aims to balance economic advancement with the resolution of critical social problems. This vision inherently demands a focus on sustainability and strength against unpredictable threats. The foundation for evaluating this balanced progress in smart cities is often the Triple Bottom Line (TBL) Theory, which emphasizes that sustainable development must equally address environmental, economic, and social goals. Environmental sustainability involves managing natural resources responsibly and reducing harmful impacts like pollution and excessive energy use. Economic sustainability requires achieving long-term economic growth while minimizing environmental harm and maximizing resource efficiency. Social sustainability focuses on enhancing the quality of life for all citizens, ensuring equitable access to resources, and reducing social inequalities. The TBL framework enables holistic evaluation, ensuring that smart city initiatives, such as those adopted by Zurich, Oslo, and Copenhagen [4–6], do not prioritize one dimension at the expense of the others.

For instance, Copenhagen's waste-to-energy initiatives address waste management (environmental), generate energy (economical), and promote equitable access to green energy (social). Additionally, Society 5.0 aligns with the Quintuple Helix model of innovation, which incorporates the environment as a critical fifth helix alongside academia, government, industry, and civil society. This model affirms that ecological sustainability is a fundamental pillar of human well-being and technological advancement. This integration ensures that smart city solutions are developed in harmony with nature. It directly contributes to global sustainability targets, like the UN Sustainable Development Goals (SDGs). In essence, the idea of Society 5.0 places human and ecological principles at the core.

4.2.2 Utilizing Advanced Technology for Smart City Development

The technological backbone of the smart city revolves around the intelligent application of advanced systems, such as AI and DT, to enhance urban resilience and efficiency. AI, along with IoT and big data, plays a transformative role in advancing environmental goals by optimizing resource use, managing waste, and improving urban transportation systems. AI's potential extends to achieving net-zero emissions through intelligent climate modeling, emissions monitoring, and renewable energy integration. For instance, AI-driven solutions are crucial for precision farming within smart cities, optimizing resources such as water utilization and chemical use to enhance food security and environmental sustainability.

A robotic crop-farming [7] prototype employing Deep Reinforcement Learning (DRL) demonstrated a 26% increase in yield, a 41% reduction in water use, and a 33% decrease in chemical use in a simulated setting. It shows the power of intelligent systems for sustainable resource management. Additionally, AI and Machine Learning algorithms analyze vast amounts of data generated by interconnected systems (IoT and sensors) to provide valuable insights, predict outcomes, and optimize complex processes. Digital Twin technology is emerging as a key orchestrator in this ecosystem, creating dynamic, virtual representations of urban assets, processes, and systems. In an urban context, a digital twin technology can enable innovative city management, such as modeling pollution or simulating the impact of a new park on local temperatures and pollution levels. Society 5.0 emphasizes that DTs and other AI models must be transparent and publicly auditable, requiring a Continuous Validation, Verification, and Accreditation process to ensure accuracy and alignment with societal goals. This continuous validation ensures urban systems are strong, adaptive, and trustworthy, which is vital for policymaking and management in Society 5.0.

4.2.3 Optimizing Resource Flows

Optimizing resources is an essential application of smart city frameworks, specifically those aligned with environmental sustainability, concerning energy consumption, waste management, and air quality. In terms of energy management, Zurich's ambitious [4] "2000-watt Society strategy" aims to reduce per-person per-year energy consumption to 2000 W and CO_2 emissions to 1 tonne. It includes promoting renewable energy adoption and using green spaces for natural cooling, reducing the need for artificial climate control. Oslo [5] emphasizes transitioning to renewable energy sources, which account for 60% of its total energy consumption, primarily from hydro-power, alongside investments in electric mobility to reduce reliance on fossil fuels. Copenhagen [6] implements smart energy grids to optimize energy use and integrate renewable sources, resulting in a 10% reduction in global warming. Regarding waste management, cities aim for streamlined, efficient, and green processes. Zurich has improved waste management practices and recycling through the integration of Information and Communication Technology (ICT). Oslo uses big data for urban planning to support more efficient waste collection, contributing to a cleaner city. Copenhagen utilizes innovative approaches, such as waste-to-energy programs that convert waste into usable energy, which contribute to its ambitious carbon-neutrality goals. However, some smart city solutions, such as pneumatic waste collection systems in Copenhagen, have had mixed results, sometimes increasing the environmental burden. Zurich reduces transportation-related emissions, which account for around 40% of total greenhouse gas (GHG) emissions, by encouraging public transport, cycling, and walking. Oslo's promotion of electric transport and reduced car usage has led to measurable improvements, including a 22% decrease in Particulate Matter (PM) and Nitrogen Oxide (NO) emissions since 2016. Copenhagen also tracks air pollution levels using data-driven solutions like sensor networks, enabling proactive management and greener urban environments. These efforts highlight how data and integrated policies are utilized to refine resource consumption and achieve specific climate goals.

4.2.4 Human-Centric Urban Planning

The Society 5.0 framework insists that smart city development must be rooted in human-centric principles, such as improving quality of life, fostering unity, and ensuring equitable access to resources. There is a need to plan proactively to reduce social harm from technological advancements and societal growth. In smart cities like Zurich, Oslo, and Copenhagen, adopting a human-centric lens means actively involving citizens and prioritizing public benefits. Zurich practices cooperative urban planning and stakeholder engagement. Oslo, recognized as a leader in sustainable urban development, faces challenges in ensuring equitable access to green innovations for all socioeconomic groups. Copenhagen, known for its focus on social inclusion, emphasizes citizen involvement in decision-making processes. Alongside this, the Kochi Area Redevelopment Project [8] aims to revitalize the city's heritage areas into a sustainable, people-centered tourism zone. It prioritizes community participation, walkable streets, waterfront promenades, and pedestrian-friendly public spaces. The plan also enhances the Mattancherry precinct with parks, bazaars, and local economic areas, improving residents' quality of life while preserving the city's urban character.

The commitment to social equity requires addressing potential disparities, as highlighted by the World Cities Report 2016, which calls for equal access to resources, social upliftment, and disaster management. Furthermore, the paradigm shift toward a highly networked, data-driven society requires a foundation of trust and transparency. Public disbelief and distrust toward data-driven initiatives, particularly in cities like Oslo, represent significant obstacles that inclusive governance and communication must overcome. Technology must empower citizens, so widespread scientific literacy and computational thinking are essential. It enables citizens to validate the solutions and actively work with complex digital systems. Integrating safety, well-being, and social sustainability is vital for creating strong, inclusive smart cities.

4.3 Industry 5.0: Cobot

Industry 5.0 marks a significant transition in manufacturing. It moves beyond the technology-driven model of Industry 4.0 and introduces a value-oriented perspective. This phase focuses on meaningful collaboration between humans and advanced technologies. Its core principles highlight human-centric design, flexibility, and sustainability.

While Industry 4.0 prioritized automation, efficiency, and smart systems, Industry 5.0 advances these ideas with a stronger social and ecological purpose. It acknowledges that human abilities, such as creativity, critical thinking, and emotional intelligence, remain essential, even in highly technological environments. By combining human strengths with intelligent machines, Industry 5.0 [9] seeks to build production systems that are adaptable, flexible, and strong. Ultimately, this approach supports not only industrial progress but also broader societal well-being.

4.3.1 Shift from Technology to Value

Industry 5.0 is primarily distinguished from Industry 4.0 by its focus on value rather than technology. While Industry 4.0 focused on integrating digital technologies to automate and optimize processes, Industry 5.0 prioritizes people, sustainability, and social benefits. This evolution explicitly shifts the manufacturing [10] focus from maximizing efficiency alone to promoting human well-being and recognizing the critical role of human expertise. Humans and machines must collaborate under Industry 5.0, as automation [11] cannot replicate essential human skills such as creativity, critical thinking, problem-solving, and emotional intelligence. Technology, therefore, is utilized to improve human capabilities rather than replace them entirely.

A notable example of this approach is the use of collaborative robots (cobots), which are designed to work alongside human operators. They assist in tasks that require precision, strength, or high repetition. By automating physically demanding tasks, Industry 5.0 [12] significantly improves employee well-being and workplace safety, enhancing the overall quality of life at work. This shift allows workers to focus on creative and intellectual tasks. It encourages higher job satisfaction and retention.

Furthermore, Industry 5.0 emphasizes employee skill development. In this collaborative environment, human abilities such as adaptability, empathy, and critical thinking become essential. As a response, companies are investing in training and development to strengthen these skills. It prepares workers for roles that continue to evolve as technology advances. This human-centric approach ensures that technologies work as a copilot alongside humans. It transforms the industrial landscape into a collaborative and personalized ecosystem. The outcome is a flexible, resilient production environment that acknowledges both human input and technological advancements.

4.3.2 Embedding Sustainability and the Circular Economy

Sustainability is a non-negotiable pillar of the Industry 5.0 paradigm. It aims to align industrial progress with global sustainability and societal benefits. Industry 5.0 adopts resource-efficient methods while embedding sustainability into all manufacturing processes. Embedding sustainability and the circular economy promotes environmentally responsible practices as a core function of systems, products, and organizations. Instead of using resources once and discarding them, the focus shifts to reducing waste, reusing materials, and designing products that can be recycled or repurposed. This approach is beneficial for long-term environmental balance.

This framework explicitly supports global targets, such as the Sustainable Development Goals (SDGs). The growing use of AI, IoT, and Big Data helps optimize resources and support sustainability requirements. For example, AI algorithms can analyze data to identify inefficiencies in the supply chain process and suggest improvements. AI improves logistics by forecasting demand, reducing extra stock, and planning efficient routes. It helps cut emissions and lower fuel expenses.

Additionally, Distributed Ledger Technology, such as blockchain, improves transparency and tracking in supply chains. It ensures that suppliers comply with environmental regulations and sustainable practices. Industry 5.0 supports the circular economy by reducing waste and promoting recycling, reuse, and regeneration. It creates complex parts with minimal waste and supports on-demand, local production. It reduces lead times and removes the need for large inventories. 3D printing enables the creation of lighter operational parts, reducing energy consumption.

4.3.3 Adaptive Manufacturing Systems

A key idea in Industry 5.0 is building robust, adaptable systems capable of handling challenges and disruptions. The COVID-19 pandemic highlighted the need for resilient supply chain systems and flexible industrial practices, prompting industries to adopt responsive approaches.

Industry 5.0 provides a roadmap to this goal by enabling flexible, human-centric, and sustainable production processes. The collaborative nature of Industry 5.0 enables flexible and agile manufacturing practices that quickly respond to market shifts, changing demand, and product variations. Technologies like DTs and AI support this. DT technology creates virtual versions of real systems. They help organizations test scenarios, improve processes in real time, find bottlenecks, and predict equipment failures before they happen. This predictive capability is key to enhancing resilience by reducing downtime and ensuring optimal operational performance through predictive maintenance. Human judgment remains essential for resilience. AI can analyze data and offer valuable insights, but humans draw on experience, intuition, and creativity to make strategic choices, especially in unexpected situations. Eventually, Industry 5.0 transforms industries into adaptive, innovative, and human-focused ecosystems prepared to withstand major disruptions.

4.3.4 Technology as an Enabler

Industry 5.0 uses existing technologies not to replace people, but to support their skills and improve collaboration between humans and machines. This technology-as-an-enabler approach aims to create seamless and symbiotic relationships between people and intelligent computer systems. A key application is the use of Collaborative Robots (Cobots), which are designed to work with humans. Cobots assist in repetitive or physically demanding tasks, allowing people to focus on complex and creative work. These cobots work smoothly with humans and can respond to simple gestures or spoken instructions. Digital Twin (DT) acts as the central control system in Industry 5.0. They create virtual, real-time models of physical systems using data collected from IoT sensors. It uses AI and simulations to test scenarios and support decision-making. People can monitor and guide these virtual models.

Augmented Reality (AR) and Virtual Reality (VR) play a crucial role by offering clear, immersive ways for people to interact with machines. For workers, AR places real-time instructions or data directly on equipment, helping reduce mistakes and improve efficiency. VR provides safe, realistic training environments, allowing workers to practice tasks and build skills without the risks of the real world. Finally, AI-Augmented Decision-Making systems support humans by pairing AI's analytical capabilities with human expertise. This collaborative structure integrates the "Human-In-The-Loop" (HITL) approach, in which humans oversee automated systems and assist with decision-making.

4.4 Agriculture 5.0

Agriculture 5.0 is the idea used to address the significant global challenges. As predicted in [13], the world population is expected to reach 10 billion by 2050, and food systems will face increasing stress. Climate change, soil degradation, water shortages, and pollution from traditional farming methods add to this stress. Agriculture 5.0 integrates advanced technologies as a precision farming framework to achieve greater efficiency and sustainability. It is mainly relevant for urban environments seeking food security. The robotic farming system uses Deep Reinforcement Learning to manage seeding, watering, and crop care based on real-time sensor data. Testing revealed a tangible environmental benefit. The studies show how autonomous robots can improve sustainability and resource optimization through data-based decisions.

4.4.1 Technology and Resource Optimization

Agriculture 5.0 uses tools such as Deep Reinforcement Learning (DRL), Geographical Positioning Systems (GPSs), Computer Vision, Sensors, and Data Analytics to support autonomous, data-driven decisions, rather than relying solely on human judgment. This method allows close monitoring of crops and the application of treatments as needed, improving productivity and sustainability.

A practical example of this technology is the Agro-robotic system, which demonstrates how modern farming can optimize resources and boost productivity. The system significantly reduced chemical usage. Water consumption also decreased, addressing critical concerns about resource conservation. Additionally, precise management of seed sowing and irrigation increased crop yield, as measured by total shoot length, thereby enhancing overall food production. These outcomes highlight the potential of autonomous agricultural robots and AI to make farming resource-efficient and adaptive. Such innovations lay the foundation for a new generation of intelligent agricultural systems capable of meeting the demands of a growing global population while minimizing environmental impact.

4.4.2 *Efficient Land Management*

Efficient land management is essential because agriculture faces stress from shrinking farmland, land degradation, and urbanization. As cities grow, less space is available for farming, making it necessary to use land wisely and reduce environmental harm. The main goal is to make the best use of limited land through smart and sustainable methods. Instead of relying solely on human judgment, farmers now use real-time data to make decisions. For instance, the Agro-robotic system shows how automation can improve fertilizer distribution, water use, and crop care. These improvements help reduce waste, increase efficiency, and lower the amount of land needed to produce food. Land management in Society 5.0 [14] also includes urban planning. Technologies such as Digital Twin (DT) allow cities to model and manage land use effectively. By processing data about soil, terrain, climate, and population patterns, DT helps create long-term plans that support sustainable development. They also help reduce the environmental effects of rapid population growth. Together, these methods ensure that land is used productively for food while also supporting sustainable and resilient cities. In Society 5.0, land is not only a resource for farming but also a key part of creating a balanced, human-centered future.

4.4.3 *Foster Social Well-Being*

Society 5.0 is a human-focused, highly connected, super smart society. It aims to support economic growth and address social challenges by closely linking cyberspace with physical space. In the agricultural sector, adopting Society 5.0 principles means employing AI-enabled intelligent precision farming. This technological shift is crucial for addressing significant global challenges, notably the pressure to meet the needs of an ever-increasing global population. A key requirement is to enhance food security and nutrition, particularly in growing smart cities where cultivation is constrained. Precision farming does more than save water or increase yields. It supports community learning and shared responsibility. It creates trust, inclusion, and innovation, and also strengthens social bonds. In the process, it improves well-being and strengthens urban food systems.

The human-centric focus of Society 5.0 is designed to overcome long-term societal challenges. The drive towards social sustainability aims to improve citizens' quality of life and ensure equal access to resources. By minimizing agriculture's environmental footprint and increasing productivity, Society 5.0 supports better ecological outcomes. Thus, improving public health in urban areas aligns agricultural efficiency directly with societal prosperity.

4.4.4 Redefined Labor Roles in Agriculture

The move toward Society 5.0 in agriculture brings a subsequent change in how people work. It focuses on humans first. Technology supports workers instead of replacing them. In this new era, people and machines work together. Human creativity and skills guide the process. Smart tools add speed and accuracy. Together, they create a balanced and efficient farming system.

The core objective is to shift repetitive, monotonous, or physically demanding tasks to autonomous systems. For example, the Agro-robotic robot uses Deep Reinforcement Learning for decision-making. It sows seeds and manages water using real-time sensor data. Automation reduces workplace risks and supports worker health and well-being. As a result, human labor shifts to creative and thoughtful roles. Workers handle tasks such as planning, operating robots, and analyzing data. Decision-making becomes a collaborative process that combines human intuition and expertise with AI's analytical capabilities. This shift emphasizes skills such as creativity, critical thinking, and emotional intelligence. In Society 5.0, technology supports people and helps them grow in the agricultural system.

4.5 Medical 5.0

Healthcare 5.0, also known as Medical 5.0, represents a significant shift toward smarter, human-centered care. This shift is driven by the rising number of chronic diseases worldwide, necessitating efficient, accessible, and personalized healthcare. Unlike traditional approaches, Medical 5.0 goes beyond general treatments and focuses on preventive and predictive care [15].

4.5.1 AI Integration in Healthcare

Medical 5.0 leverages advanced technologies such as AI, IoT, 5G, big data, cloud computing, and blockchain. These tools are often combined in the Internet of Medical Things (IoMT) to improve healthcare [16] services. The Learn-Predict-Monitor-Detect-Correct (LPMDC) framework provides the first comprehensive theoretical foundation for the systematic integration of AI into Critical Care/Intensive Care Units (ICUs). The LPMDC framework operates as a five-phase architecture that enables processing beyond traditional human cognitive limits through continuous feedback loops and system optimization. It improves predictive analytics, patient monitoring, smart alert, and treatment decision support while maintaining essential human clinical oversight.

Clinical studies [4] show clear benefits from LPMDC-enabled systems. Mortality rates dropped by 30% [17], demonstrating improved patient outcomes. ICU efficiency [18] has also improved, with the average length of stay reduced from 7.5 to 6.1 days and better compliance with sepsis care protocols [19]. Machine learning tools provided strong predictive capabilities, detecting sepsis with 80% sensitivity up to 3 h before symptoms appeared. Additionally, AI-assisted tools helped reduce clinicians' mental burden by organizing patient data and highlighting critical information. This support is valuable during complicated or high-stress situations. Overall, these results show that LPMDC systems can enhance patient care, improve hospital efficiency, and support healthcare professionals in making faster decisions.

4.5.2 Personalized Care and Enhanced Accessibility

A primary goal of Society 5.0 is to use technology to improve the lives of everyone. In healthcare, this goal leads to personalized, efficient, and accessible medical services. Now, the focus is on disease prevention and treatment, and personalized treatment becomes a core idea. Healthcare moves away from a "one-size-fits-all" model. Instead, treatment plans are designed based on a person's genetic information, lifestyle, medical history, and real-time health data. Tools such as remote monitoring devices and wearable sensors help doctors track patient health outside hospitals. It makes care accurate and proactive, and patients receive support at home.

With the invention of robotics and telemedicine, healthcare has become accessible globally. Telemedicine and digital platforms allow patients in rural or remote areas to connect with doctors through video calls or mobile apps. It reduces travel time, saves time, and ensures that people receive timely medical attention. For communities with limited medical facilities, these tools create equal access and better health outcomes. Digital records, automated scheduling systems, and AI-driven support tools reduce administrative work. It helps doctors and nurses spend less time on paperwork and more time attending to patients. It reduces stress and smooths the healthcare system's operations. Overall, Healthcare 5.0 aims to create a smart, fair system that focuses on human well-being. It seeks to make healthcare available to all and effective. With technology working alongside people, Society 5.0 builds a future where healthcare supports every individual.

4.5.3 Data Security for Societal Reliance

The success of the healthcare system depends on secure data systems. It also depends on the ability to stay stable during unexpected events such as pandemics, disasters, or system failures. This idea aligns with the purpose of Society 5.0, which aims to create smart systems that protect human well-being. Data plays a central role in modern healthcare. Hospitals, devices, apps, and healthcare platforms continuously collect vast amounts of information. It includes medical history, test results, movement patterns, and real-time body signals. Since this data is sensitive, protecting it is extremely important. Without trust in data security, people will hesitate to use digital healthcare services.

Privacy protection becomes a top priority in the growing medical field. Health information must remain confidential and must not be misused, leaked, or accessed without permission. For this reason, healthcare relies on advanced security methods. Federated learning is one such technique for protecting data; in this approach, information does not need to be stored or shared centrally. Instead, the data stays where it is collected, such as on a device or in a local hospital system. Machine learning models are trained by sharing patterns, not the patient information. It allows technology to improve without exposing personal data. This approach is instrumental in healthcare, where strict rules protect patient information. Federated Learning enables cooperation among hospitals, research labs, and countries without compromising privacy. It also helps during emergencies when fast data analysis is needed, but sharing sensitive information is unsafe.

4.5.4 Responsible Use of Digital Twin Technology

In Medicine, Digital Twin technology plays an essential role in refining patient care. A Digital Twin is a virtual replica of a patient or system that can be used to test treatments before they are prescribed in real life. In critical care environments such as ICUs, this technology helps simulate different medical responses and predict outcomes. However, because patient safety is involved, these models must undergo rigorous processes known as Verification, Validation, and Uncertainty Quantification (VVUQ). These steps ensure that the predictions closely match real clinical data and can be trusted by medical professionals. Since healthcare involves life-and-death decisions, strong ethical standards are essential. Every AI system must follow accountability principles to ensure its recommendations are transparent and easy for clinicians to understand. Doctors need to see the clear reasoning behind the system's suggestions, rather than accepting outcomes unquestioningly. Regular audits are also required to prevent bias in AI systems, especially toward different age groups, genders, or ethnic backgrounds. These checks ensure fairness and help maintain trust in digital healthcare tools.

It is also crucial to prioritize human judgment. AI should support decision-making, and not replace physicians. The ultimate responsibility must always rest with trained medical professionals who understand the full context of patient care. Looking ahead, Medical 5.0 will continue to evolve as technology advances. Quantum computing may enable computers to process medical data much faster. Federated learning will allow the use of shared data models without exposing patients' private information. Explainable AI will help ensure that technology remains transparent. Together, these innovations will help create a safer, smarter, and human-centered healthcare system. Figure 4.1 demonstrates the application of Society 5.0.

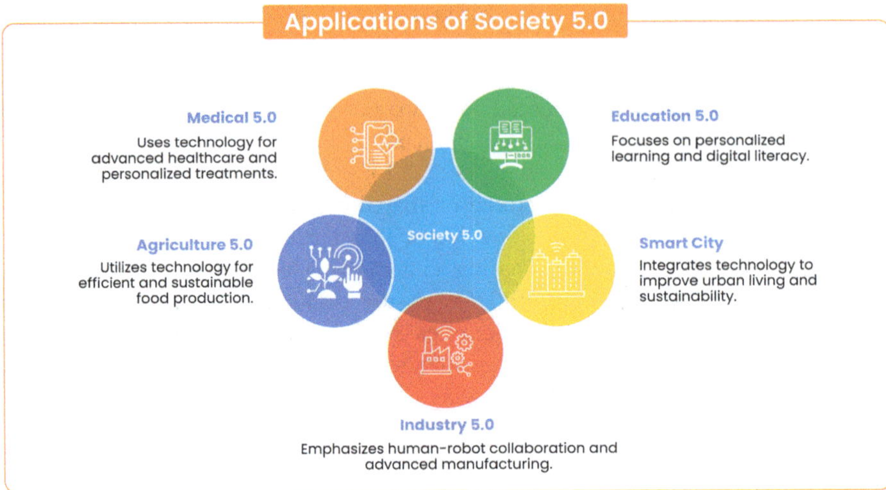

Fig. 4.1 Application of Society 5.0

4.6 Conclusion

Society 5.0 represents a systematic shift in how societies manage human life, moving beyond economic efficiency to address major social problems. Unlike previous technological revolutions that focused predominantly on efficiency and economic growth, Society 5.0 is a prospective, policy-driven construct that uses advanced technologies such as AI, IoT, big data, and robotics to achieve inclusivity, sustainability, and resilience. This strategy applies across all the essential sectors. In agriculture, the shift to Agriculture 5.0 addresses food shortages, severe labor gaps on farms, and the impacts of a changing climate. It uses tools such as cobots and digital twins to maximize production while integrating human intelligence into critical decision-making. This approach also allows farmers to customize products to consumer preferences, offering developing nations, such as Nepal, the opportunity to skip older industrial stages and modernize rapidly. Beyond the farm, the applications revolutionize healthcare by using AI to help doctors deliver personalized medicine, monitor patients remotely, and predict health issues before they become crises, shifting the focus from treating illness to maintaining wellness. Education also sees a profound change, with advanced learning programs tailoring content in real time to individual students. Virtual reality provides hands-on training and simulations, significantly improving access to learning, particularly for students in remote areas. Even business operations, such as human resources, integrate these technologies to meet environmental goals, reducing waste and encouraging staff to adopt sustainable behaviors. This profound integration promises substantial improvements in citizens' daily lives.

Despite the clear benefits, realizing this vision requires that societies actively confront several significant threats that could prevent universal welfare. The primary risk remains the gap between those who can access technology and those who cannot, known as the Digital Divide. This uneven distribution of skills and tools risks exacerbating social and economic discrimination, affecting income, age, and geography. Furthermore, the reliance on high-automation technologies, such as robots and AI, poses an immediate threat to jobs and stability, creating severe economic anxiety and psychological strain for workers who lack opportunities to retrain. Beyond these social challenges, fundamental ethical and legal issues demand immediate attention. If the algorithms that run these smart systems are trained on biased or unrepresentative historical data, they will inevitably perpetuate and intensify existing social prejudices.

While collecting extensive personal data enables a seamless cyber-physical experience, it also brings important privacy and civil-liberty considerations that must be carefully addressed. To safeguard citizens, governments must establish strong data protection laws, build resilient cybersecurity frameworks, and ensure completely transparent governance. Additional hurdles include the substantial initial costs for companies and local authorities to adopt these complex technologies, the difficulty in creating flexible laws to govern mass customization, and the pressing need for continuous investment in research and development to close critical skill gaps across the workforce. The success of this human-centered society ultimately depends not on technical capability alone but on intentional strategic investment, comprehensive educational programs, and genuine collaboration among all stakeholders to ensure innovation is ethical, inclusive, and environmentally sound.

References

1. Nieto-Taborda, M. L., & Luppicini, R. (2024). Accelerated digital transformation of higher education in the wake of COVID-19: A systematic literature review. *International Journal of Changes in Education, 2*. https://doi.org/10.47852/bonviewIJCE42023125
2. Borra, S., Dey, N., & Peng, S. L. (2026). Human-centric AIoT strategies for sustainable business evolution. In *AIoT deployment in business evolution: Sociotechnical readiness and human-centric strategies* (pp. 47–57). Springer Nature Switzerland.
3. Oyadeyi, O. A., & Oyadeyi, O. O. (2025). Towards inclusive and sustainable strategies in smart cities: A comparative analysis of Zurich, Oslo, and Copenhagen. *Research in Globalization, 10*, 100271.
4. 2000-Watt Society. (n.d.). Worldholder@Work. Retrieved from https://www.2000-watt-society.org/worldholderwork?utm_source=chatgpt.com
5. Modeshift. (n.d.). Oslo is ranked one of the most sustainable cities in the world here's why. Retrieved from https://www.modeshift.com/oslo-is-ranked-one-of-the-most-sustainable-cities-in-the-world-heres-why
6. Lund, H., Thellufsen, J. Z., Sorknæs, P., Mathiesen, B. V., Chang, M., Madsen, P. T., et al. (2022). Smart energy Denmark. A consistent and detailed strategy for a fully decarbonized society. *Renewable and Sustainable Energy Reviews, 168*, 112777.
7. Addas, A., Tahir, M., & Ismat, N. (2024). Enhancing precision of crop farming towards smart cities: An application of artificial intelligence. *Sustainability, 16*, 355.
8. Bee Smart City. (2021, May 25). Smart City Kochi—A vibrant and inclusive city of opportunities. Retrieved from https://www.beesmart.city/en/smart-city-blog/smart-city-kochi-india

9. Khosravy, M., Gupta, N., Pasquali, A., Dey, N., Crespo, R. G., & Witkowski, O. (2023). Human-collaborative artificial intelligence along with social values in industry 5.0: A survey of the state-of-the-art. *IEEE Transactions on Cognitive and Developmental Systems, 16*(1), 165–176.
10. Deshpande, R. (2025). How Manufacturing 5.0 in Telangana can accelerate industrial transition.
11. Adel, A. (2023). Unlocking the future: Fostering human–machine collaboration and driving intelligent automation through Industry 5.0 in smart cities. *Smart Cities, 6*, 2742–2782.
12. Rejeb, A., Rejeb, K., Zrelli, I., Kayikci, Y., & Hassoun, A. (2024). The research landscape of Industry 5.0: A scientific mapping based on bibliometric and topic modelling techniques. *Flexible Services and Manufacturing Journal, 37*, 1203–1250.
13. United Nations, Department of Economic and Social Affairs, Population Division. (2017). *World population prospects: The 2017 revision*. United Nations. Retrieved from https://www.un.org/development/desa/pd/content/world-population-prospects-2017
14. Fonseca i Casas, P., & Pi i Palomes, X. (2026). Building Society 5.0: A foundation for decision-making based on open models and digital twins. *Advanced Engineering Informatics, 69*, 103970.
15. Patil, R. V., Ambritta, N. P., Mahalle, P. N., & Dey, N. (2022). Medical cyber–physical systems in Society 5.0: Are we ready? *IEEE Transactions on Technology and Society, 3*(3), 189–198.
16. Rahmouni, H. B., Hassine, N. B. E. H., Chouchen, M., Ceylan, H. İ., Muntean, R. I., Bragazzi, N. L., & Dergaa, I. (2025). Healthcare 5.0-driven clinical intelligence: The learn-predict-monitor-detect-correct framework for systematic artificial intelligence integration in critical care. *Healthcare, 13*, 2553.
17. Institute for Health Metrics and Evaluation. (2025, October 11). New Global Burden of Disease study: Mortality declines globally, but youth deaths surge in stark health divide [Press release].
18. Ansari, N., Mittal, P., Ajmera, P., & Dadhich, A. (2024). Fundamentals of Healthcare 5.0.
19. Alanazi, A., Aldakhil, L., Aldhoayan, M., & Aldosari, B. (2023). Machine learning for early prediction of sepsis in intensive care unit (ICU) patients. *Journal of Healthcare Engineering, 2023*, 6616453. https://doi.org/10.1155/2023/6616453

Chapter 5
Opportunities and Challenges in Society 5.0

5.1 Introduction: The Dawn of Society 5.0

Society stands at a critical inflection point. Beyond efficiency, it reshapes technology to serve humanity meaningfully. Society 5.0, first developed in Japan, is not just another digital step but a new vision that aims to balance economic growth with the resolution of major social problems, such as aging populations, limited resources, and inequality. While Industry 4.0 focused almost exclusively on optimizing manufacturing and supply chains, Society 5.0 places the human at the center. The core technological engine driving this transformation is the Cyber-Physical System (CPS). It is not merely the Internet of Things (IoT), but a deeply integrated network in which sensors, actuators, computational models, and human interactions blend seamlessly. Data collected in the physical world is analyzed in cyberspace by sophisticated AI tools, and the resulting insights are immediately applied back to the physical world to initiate corrective action. This change brings great opportunities but with significant risks, and handling it wisely is the generation's primary challenge.

5.1.1 The Paradigm Shift

The shift from Industry 4.0 to Society 5.0 represents a fundamental transition from a purely economic goal to human-centric societal needs. Industry 4.0's primary focus was the creation of "smart factories" to maximize productivity and lower costs through automation and data exchange. While this delivered greater efficiency, it often disregarded the social conflicts (such as job displacement and ethical data use) that technology created. Society 5.0 repurposes advanced technology to solve challenges and support humanity. The goal is to create a "Super-Smart Society"

S. Shukla et al., *Society 5.0*, SpringerBriefs in Applied Sciences and Technology, https://doi.org/10.1007/978-3-032-22164-3_5

where technology integrates into every facet of human life. It ensures that data-driven solutions benefit all citizens, not just the privileged. This change makes technology support sustainability, inclusion, and improved quality of life. This shift requires a re-evaluation of national priorities, moving from simply chasing economic growth to promoting human security and development.

5.1.2 The Core Technology: The Cyber-Physical System (CPS)

The foundational architecture enabling Society 5.0 is the Cyber-Physical System (CPS). It is a complex, closed-loop technical architecture where physical processes are intensely monitored and controlled by embedded computation and communication networks. Unlike traditional IT, the CPS demands low-latency and high-reliability operations, as system failures can have immediate damage (e.g., power grid failures or medical device malfunctions). The system operates through three core stages: sensing (data acquisition from biosensors), computation (AI and machine learning algorithms for analyzing and modeling the data), and actuation (execution of decisions in the physical world via robotic systems, smart meters, or autonomous vehicles). Stochastic processes, which take into account the variability and uncertainty of the real-world environment, are used to represent the mathematical framework that provides reliable control. The CPS acts as the nervous system of Society 5.0, translating billions of data points into actionable insights that automate and optimize human life.

5.1.3 Balancing Opportunities and Challenges

The central idea behind the success of Society 5.0 depends on carefully balancing advanced technologies with serious ethical and practical challenges. The integration of the CPS brings remarkable advancements in personalized health, environmental efficiency, and economic agility. With the benefits come risks: bias, inequality, and reduced control over data. It highlights transformative technologies and supports them with real cases. These advancements require a responsible way forward, emphasizing agile governance and sustained human-centric development. For example, in Society 5.0, a smart hospital uses AI to provide personalized life-saving treatments. However, to prevent discrimination, human auditors must continuously check these algorithms for bias. This balance ensures that while technology accelerates healthcare efficiency, it remains fundamentally fair and transparent for all patients.

5.2 The Opportunities

Figure 5.1 presents the opportunity in Society 5.0.

5.2.1 *Transforming Healthcare*

One of the most essential benefits of Society 5.0 is the revolution in healthcare, mainly addressing the worldwide challenge like aging populations. The future of healthcare moves away from reactive treatment toward preventive care driven by continuous data monitoring. Society is entering the era of precision health, fueled by the convergence of AI, genomics, and advanced biosensors. Instead of prescribing treatments based on population averages, physicians now utilize AI to analyze an individual's unique clinical data alongside lifestyle metrics collected via wearable devices. It allows for truly personalized treatment plans. This technological shift extends to caregiving, with the development of assisted-living robots that help with mobility and medication reminders.

5.2.1.1 The Era of Precision Health

The combination of genomics and proteomics is ushering in a new era in which healthcare is moving toward personalized treatments based on each individual's genetic and protein information. Instead of using one treatment for all, this approach designs a treatment plan and therapies that match each person's unique molecular

Fig. 5.1 Opportunities in Society 5.0

profile. In this scenario, AI and machine learning algorithms are indispensable tools for processing petabytes of data generated by advanced molecular sequencing. The AI models identify subtle disease biomarkers and gene-drug interactions that are invisible to the human eye, predicting a patient's likely response to a specific treatment. They cut both the time and cost of developing treatment plans and improved patients' outcomes by avoiding traditional trial-and-error methods. The increase in healing precision minimizes adverse effects and maximizes the efficacy of expensive treatments, making healthcare sustainable and effective. For example, AI-based genomics combined with remote biosensors for the management of chronic diseases. Patients wear smart patches that continuously track body data, which is compared to a DNA-based digital twin. AI detects early risk signals and proactively adjusts treatment, enabling personalized care before symptoms develop. The company named "Tempus" [1] already uses this technology to customize chemotherapy, shifting healthcare from reactive hospital visits to proactive, personalized treatment that extends healthy life expectancy.

5.2.1.2 Remote Monitoring and Wearable Diagnostics for Preventive Care

The invention of highly accurate, noninvasive biosensors has driven the paradigm shift from treating illness to sustaining wellness. These sophisticated devices are embedded in clothing, smartwatches, or skin patches to continuously monitor physiological data, including heart rate variability, blood glucose levels, and respiration. The collected data provides a continuous stream of health status, which is analyzed by edge-computing AI systems. It enables the timely detection of anomalies that precede a medical crisis, facilitating genuine preventive care. For example, in oncology, companies such as Tempus and Foundation Medicine use deep learning to analyze millions of anonymized tumor samples and corresponding genetic markers. They allow oncologists to predict how a patient will respond to a specific chemotherapy agent with high accuracy, moving beyond trial and error. Additionally, advanced biosensors can alert caregivers or emergency services to subtle changes in a patient's vital signs, a significant drop in blood oxygen, or an irregular heart rhythm before a critical event occurs. This approach shifts chronic disease treatment toward extending healthy living, not just adding more years to life.

5.2.1.3 Robotics in Caregiving

Addressing the severe global shortage of human caregivers, Society 5.0 promotes the development of assisted living robotics. These robots are designed not to replace human warmth but to enhance human capability and capacity in demanding physical tasks, thereby allowing human caregivers to focus on emotional and social interaction. Patients regain freedom of movement with exoskeletons, while companion robots provide mental stimulation and social companionship. Ethically, organizations should develop the technology to preserve the dignity and autonomy of older people, not reduce the quality of human interaction. Robotics takes over the strenuous, repetitive, or potentially dangerous aspects of care, such as lifting patients, monitoring sleep patterns, or reminding people to take their medications on time. Technology handles routine tasks so that caregivers can spend more time on emotional support, empathy, and complex decision-making. For example [2], PARO is a well-known robot used in assisted living for elderly people. This cute robotic seal helps in elderly care by giving comfort and reducing stress. It reacts to touch, light, sound, and posture, making patients feel calm and happy. For those with dementia or loneliness, PARO boosts brain activity, enhances mood, and encourages social interactions. Studies show it reduces anxiety and medication needs. Easy to use and safe, PARO demonstrates how robotics may improve the effectiveness and comfort of senior care.

5.2.1.4 Data-Driven Wellness

Using aggregated, anonymous health information, Society 5.0 enhances both individual wellness and community-wide public health management. The massive collection of data from clinics, hospitals, and personal sensors powers AI models for forecasting health crises. These models can anticipate the spread of infectious diseases (e.g., pandemics), predict spikes in chronic disease cases (e.g., asthma attacks triggered by local air quality), or identify geographic areas with unmet healthcare needs. These indicators allow public health officials to strategically allocate resources, deploy mobile clinics, and launch targeted precautionary campaigns (e.g., awareness camps and vaccination drives). By using data-driven forecasts, public health systems can minimize waste, mitigate risks early, and strategically deploy resources to achieve the most significant public benefit. This capability is essential for creating a highly resilient and responsive global health system capable of managing the complex challenges. For example, Johnson & Johnson shows [3] how data can support employee health. The company analyzes biometric screening, participation data, and lifestyle data to identify common problems such as high cholesterol or low physical activity. They then offer personalized support and rewards. This approach increased participation, reduced health risks, and lowered healthcare costs. It delivered an ROI of $2.71 to $3.92 per dollar spent, significantly reducing healthcare costs.

5.2.2 Smart Cities and Environmental Sustainability

Society 5.0 offers a unique way to address global environmental and resource challenges by treating cities as a single connected system rather than as separate parts. For example, in Fujisawa, Japan, smart homes, transport, energy grids, and waste systems are linked and share real-time data. When energy use rises, the system automatically balances supply and reduces waste. Residents don't notice the adjustments, but the city becomes cleaner, safer, and more efficient. This approach shows how treating a city like a living body can help protect resources while improving daily life for everyone. In a body, all parts-like the heart, brain, and lungs work together and share information. If one part needs help, the others support it. In the same way, a city can connect roads, electricity, water, transport, and waste systems, as well as homes, so they communicate and work together. When everything is connected, the city can save power, reduce waste, and make life easier for people without them needing to do anything.

5.2.2.1 Digital Twins for Smart City

Urban planning [4] in Society 5.0 moves from static models to dynamic, living virtual environments with digital twins. A digital twin is a real-time digital model that mirrors a physical building, city, or infrastructure system. These models integrate vast streams of data, including traffic flow, energy consumption, air quality sensor data, and geospatial data. It allows planners to simulate the complex interactions of the urban environment. In cities such as Singapore and Barcelona, city planners use these models to predict how new transport changes will affect traffic and air quality. They provide predictive insights into policy changes, reducing unforeseen negative consequences and enhancing positive impact. It helps city planners try to transport, waste, or service changes digitally first, ensuring the best results before actual construction begins. One of the best examples is Virtual Singapore [5], which offers a live 3D model of the whole city. They simulate urban settings, such as wind flow to minimize heat and solar potential for green energy, using real-time sensor data. It enables the city to test infrastructure improvements virtually before putting them into practice. Planners use it to model emergency evacuations and optimize barrier-free pathways for elderly.

5.2.2.2 Energy Optimization and the Smart Grid

The modernization of electrical infrastructure is critical for environmental sustainability. The existing electrical grids are outdated; these grids were designed for one-way power flow from centralized plants. The Smart Grid, however, is a digital, bi-directional network that acts as a vast, intelligent controller. It uses embedded IoT sensors and machine learning to gather and act on real-time data, enabling the system to route power dynamically, balance loads, and seamlessly integrate distributed energy resources such as rooftop solar panels and electric vehicle charging stations. This ongoing optimization reduces energy waste, enhances system reliability, promotes green energy development, and maintains a stable, reliable power supply. The aim is to build an efficient, self-healing energy system that uses green power judiciously to reach net-zero emissions. For example [6], Brooklyn, the Microgrid demonstrates smart grid optimization. Launched in 2016 by LO3 Energy, this blockchain-powered platform lets Brooklyn residents trade solar energy peer-to-peer. Smart meters track production and consumption in real-time, cutting transmission losses by 30%, lowering costs, boosting green energy, and ensuring resilience during outages like Superstorm Sandy.

5.2.2.3 Next-Generation Mobility

A unified framework called Mobility-as-a-Service (MaaS) uses connected technologies to reduce urban congestion and cut vehicle pollution in the cities. MaaS integrates various transport modes, such as public transit, ride-sharing, electric scooters, and autonomous vehicles, into a single digital platform, optimizing route planning, ticketing, and payment for end users. The foundation of this system is the Autonomous Vehicle Network, where AI-driven vehicles communicate with each other and with roadside infrastructure (Vehicle-to-Infrastructure (V2I) communication). Continuous data exchange helps vehicles travel together smoothly, reducing sudden stops and improving traffic flow for faster and safer movement. As a result, vehicles have shorter idle periods, lower fuel or electric use, and a noticeable decrease in pollution. Focusing on shared autonomous electric vehicles rather than private cars lets cities reclaim parking space and enhance urban life. In Helsinki, Whim is a cutting-edge Mobility-as-a-Service (MaaS) platform that combines bikes, taxis, and public transportation into a single digital interface. They operate as a cyber-physical system and determine the most effective multimodal routes using AI and real-time GPS data. By providing subscription-based travel, Whim promotes a smooth, inclusive, and sustainable transportation ecosystem while decreasing carbon emissions and urban traffic by shifting the emphasis from automobile ownership to human-centric outcomes.

5.2.2.4 AgriTech and Food Security

Society 5.0 extends beyond the city limits to handle global food security and resource shortages through precision agriculture. It uses a network of IoT sensors embedded in the soil, aerial imagery from drones, and machine learning models to treat individual plants or a small field, providing customized treatment. Instead of treating the entire field, farmers use data to target sections that need water, nutrients, or pest control. Using spectral analysis to monitor crops can drastically reduce water usage and significantly lower pesticide and herbicide requirements. Precision agriculture reduces resource waste, boosts yields per hectare, and protects the environment. It is a critical technological intervention essential to meeting the food demands of a growing global population amid increasingly unpredictable climate conditions. The perfect example of AgriTech in the agricultural industry is the Kubota Smart Village [7] in Japan. This technology monitors crop health using drones with spectral sensors and autonomous tractors. The technology applies fertilizer and water only where necessary, based on soil data and weather trends in cyberspace. This accuracy maximizes food security and minimizes environmental impact by reducing resource waste by up to 90%.

5.2.3 *Smarter Industry, Stronger Economy*

Society 5.0 changes the way individuals think about products and services. In the past, getting highly customized products usually meant paying a high price. Society 5.0 eliminates barriers between affordability and personalization. It focuses on flexibility and personalization, making products and services best suited to individual needs without extra cost. People no longer need to own everything; instead, they can access services as needed. This approach saves resources and reduces waste. Companies can respond quickly to changing customer demands and deliver solutions designed for specific requirements. Overall, Society 5.0 encourages a smarter, more efficient economy in which people benefit from both personalization and affordability.

5.2.3.1 Delivering Outcomes Through Services

The modern economy focuses on services, an industrial shift in which value is derived not from selling a tangible good but from guaranteeing a specific outcome or service. For instance, rather than owning a motor, a company nowadays chooses to buy guaranteed machine uptime as a service. This model is only possible because of integrated IoT sensors that continuously monitor motor performance data. Predictive Maintenance AI analyzes continuous data streams to forecast mechanical failure hours or days in advance. It helps companies to take proactive action and schedule repair. It eliminates costly, unexpected breakdowns and inventory waste. With real-time monitoring, factories can create customized single products as cheaply as mass production, transforming old manufacturing models. Rolls-Royce's "Power-by-the-Hour" strategy is a seminal example of achieving results through services in the industrial sector. Rather than only selling jet engines, the company offers a service in which airlines pay for "guaranteed flight hours." Rolls-Royce uses AI and embedded sensors to undertake predictive maintenance by continuously monitoring engine status. By optimizing resource efficiency, minimizing unplanned downtime, and establishing a sustainable, data-driven value chain, this transition from selling a product to guaranteeing an outcome is consistent with Society 5.0.

5.2.3.2 FinTech: Future of Finance

FinTech revitalized innovation in the global financial system through Distributed Ledger Technology (DLT), primarily blockchain. DLT provides a trusted and permanent record of transactions. As a result, it reduces fraud and makes both banking and supply chain processes more efficient. This technology has a significant impact on people without bank access. It can help billions of people who currently lack access to formal financial systems. By utilizing secure digital identity and mobile access to decentralized finance platforms, Society 5.0 brings financial services to remote communities, facilitating safe, low-fee, cross-border transactions and microloans.

Furthermore, the use of biometric authentication and digital transaction systems significantly reduces the economic friction and security risks associated with cash-based economies, accelerating global economic participation and verifiable trust. M-PESA [8] is a FinTech ecosystem that prioritizes mobile devices and promotes financial inclusion for millions of people. Its AI-powered Cyber-Physical System creates real-time credit scores for the unbanked using alternative data, such as transaction history and utility payments. This human-centric method illustrates Society 5.0's vision of universal economic participation by enabling rapid microloans that overcome typical infrastructure gaps, empowering small enterprises, and reducing poverty.

5.2.3.3 Supply Chains: Visibility and Risk Forecasting

Traditional global supply chains are very complex and challenging to manage. Unexpected events like pandemics or geopolitical conflicts can easily disrupt them. With AI and the IoT, Society 5.0 builds flexible supply chains. They provide complete, real-time tracking and insight across all operations. Every component, from raw material to finished product, is tracked via sensors or digital markers, creating a continuous, auditable data trail. Machine learning analyzes the data and considers worldwide economic signals to forecast risks. It helps prevent the bullwhip effect, in which small demand shifts lead to significant inventory problems. By accurately predicting demand and supply delays, companies can use just-in-time logistics with high accuracy. They reduce extra inventory, lower waste, and build resilience against global disruptions. DHL Express [9] uses AI, IoT sensors, and robotics for real-time parcel tracking and predictive maintenance. In Japan, it integrates human workers with cobots for last-mile delivery, cutting delays by 25% and emissions through optimized drone routes. Data analytics customizes customer alerts, boosting efficiency and sustainability in supply chain and logistics.

5.2.3.4 Inclusive Startup Ecosystem

Society 5.0 provides a strong digital infrastructure and computing power. As a result, it becomes easier for individuals and startups to innovate and become entrepreneurs. With cloud platforms and free AI models available, businesses avoid heavy spending on hardware and licensed software costs, making advanced technology easier to access. This technological democratization allows small-to-medium enterprises (SMEs) to grow. They let them compete globally by using large digital platforms and their broad reach. With modular production and customization, entrepreneurs work faster. They test ideas, change designs quickly, and shorten product development from many years to just months. Such agility creates a lively startup ecosystem that replaces outdated ideas with better ones and enables many small innovators to shape economic progress. For example, Station F [10] in Paris shows how a startup ecosystem can be inclusive by reducing barriers for underrepresented entrepreneurs. Through its Founders Program and FemTech initiatives, it offers diverse founders shared access to advanced AI tools, funding, and global networks. By democratizing the platform economy, Station F supports a human-focused entrepreneurial environment that promotes economic growth through social diversity and ensures innovation benefits a wide range of society.

5.2.4 *Human Empowerment Through Lifelong Learning*

A human-centric society demands empowered citizens. In Society 5.0, technology is not just for experts or large companies. Instead, they support everyone and give equal opportunity to learn, grow, and participate. The real potential lies in providing equal access to education and essential digital tools. When learning becomes accessible to all, talent can rise from any background, not only from those with privilege. Lifelong learning systems can further support this goal. Instead of one teaching method for everyone, learning paths can adapt to each person's pace, interests, and abilities. It helps people build confidence, develop relevant skills, and stay prepared for the changing demands of work and society. Digital inclusion is also important. Reliable internet access, basic digital literacy, and access to smart tools are now as essential as electricity or clean water. When people have these foundations, they can fully take part in digital services, governance, and the economy. Ultimately, empowering citizens strengthens society. When people are informed, skilled, and confident, they can contribute ideas, solve problems, and shape a future that benefits everyone, not just the privileged ones.

5.2.4.1 Adaptive Learning Systems

The cornerstone of educational innovation in Society 5.0 is the rise of adaptive learning systems. These are advanced EdTech platforms that use AI to go beyond the limitations of same-time, one-size-fits-all education. The system analyzes learners' performance, schedules, focus times, and thinking patterns. It adjusts the lesson content, teaching pace, flexibility, and format to fit their needs. If a learner struggles with a core concept, the AI doesn't simply present a remedial exercise; it analyzes why they struggled and finds an alternative explanation or a prerequisite topic they may have missed. This tailored learning approach can finally close the gap between traditional mass education and adaptive learning. It recognizes different learning styles and allows every learner to progress at the right pace. For example, a learner who learns better through visuals may receive videos and diagrams, while another learner who prefers hands-on practice may get interactive exercises. It helps everyone gain the advanced skills needed for the modern economy. To bridge the skills gap in working professionals, LinkedIn Learning [11] uses an AI-based adaptive solution. The platform continuously customizes individual learning courses based on a user's job function, skill set, and the skill graphs of millions of similar professionals. The system monitors participation and assessment outcomes as learners interact with classes, draws real-time conclusions, and recommends customized micro-learning modules to address specific skill gaps. In Society 5.0, this people-focused approach promotes continuous learning that is applicable, useful, and directly related to professional advancement.

5.2.4.2 Augmented Reality (AR), Virtual Reality (VR), and Social Inclusion

Society 5.0 creates new opportunities for social inclusion by reducing physical and cognitive barriers with advanced technology. Tools like smart wheelchairs, voice-controlled systems, and adaptive learning platforms help people with disabilities participate fully in daily life. For example, someone with limited mobility can use voice commands to work, study, or travel independently, improving confidence and independence. AR and VR tools are transformative in this regard. AR smart glasses can provide real-time instructions or navigational assistance to workers with cognitive impairments, effectively translating the digital world into easily understandable physical cues. For individuals with mobility challenges, VR allows for participation in remote work, cultural events, and even physical therapy in a safe, simulated environment.

Furthermore, advanced assistive technology, utilizing AI-driven real-time translation and voice recognition, breaks down communication and sensory barriers for the hearing - visual impaired. This focus on technology-enabled accessibility ensures that everyone can take part in economic and cultural activities, regardless of where they live or their physical abilities. For example, remote learning platforms enable students in rural areas to access quality education. At the same time, assistive devices help people with disabilities participate in workplaces and cultural events, promoting full participation in society. For example, BMW's AR/VR [12] training program is an outstanding example of lifelong learning in a work environment when it comes immersive technologies. During assembly, workers on the production line use augmented reality glasses to superimpose digital instructions over actual engine parts, providing real-time guidance. VR simulations offer workers a risk-free virtual environment to practice complex safety procedures. The workforce can continually adapt to new manufacturing technologies without stopping production because of this high-resolution cyber-physical system, which accelerates skill development and reduces error rates.

5.2.4.3 Essential Reskilling

As machines handle routine physical and repetitive tasks, humans must focus on advanced skills. The reskilling requirements emphasize preparing workers for jobs requiring creativity, critical thinking, complex communication, and emotional intelligence. For example, while AI can process data, it cannot lead brainstorming sessions or resolve workplace conflicts, making these uniquely human skills increasingly essential. Educational and corporate training programs need to shift focus from memorizing facts to developing adaptability and problem-solving skills. For example, students or employees can work on real projects that simulate real-world challenges. This hands-on approach encourages critical thinking, teamwork, and creative solutions, better preparing learners to succeed in fast-changing workplaces and complex problem-solving environments. For the current workforce, focused upskilling programs are vital to help workers move from outdated roles to modern, in-demand jobs. For example, a data entry operator can be trained to interpret data or audit machine-learning models. Such programs ensure employees remain valuable, adapt to technological change, and succeed in evolving industries. This shift in culture and education is essential to highlight the importance of human skills in a world dominated by automation. For example, in higher-order human talents, H-FARM [13] in Italy is a prime example of an innovation platform that promotes critical thinking, creativity, and emotional intelligence.

In contrast to traditional technical training, H-FARM uses "Human-Centric Innovation" seminars to bring together experts, students, and entrepreneurs to address complex societal issues. Participants must handle varied team dynamics by navigating sophisticated communication and emotional intelligence through design thinking and collaborative problem-solving. This ecosystem demonstrates that although AI handles data, the most valuable economic outputs in Society 5.0 come from human qualities that combine technology with empathy and innovative vision.

5.2.4.4 Global Access to Knowledge

The infrastructure of Society 5.0 can make high-quality education and information accessible to everyone worldwide. For example, online learning platforms and digital libraries allow learners in remote or low-income areas to study with the same material as those in major cities. They reduce barriers and ensure equal opportunities for learning everywhere. Affordable mobile devices and broadband internet allow people in remote or developing regions to access the same quality education as urban students. For example, real-time AI translation tools remove language barriers, allowing learners to study scientific articles, academic courses, or vocational training in their native language, and ensuring equal learning opportunities for everyone regardless of location or background. This capability is crucial for accelerating global development and reducing knowledge gaps between nations. It provides opportunities that are not limited to physical borders or economic constraints. It builds an international, human-centered knowledge system where everyone can contribute and benefit equally. For example, the collaboration between Khan Academy [14] and Starlink is a revolutionary approach to breaking down geographical barriers to global access to knowledge. The system establishes physical foundation for digital knowledge by deploying low-earth-orbit satellite clusters to deliver high-speed internet to isolated areas in the Amazon and Sub-Saharan Africa. It enables learners in remote communities access to the same top-notch educational resources as those in large urban centers when paired with AI-driven translation tools. It ensures that a person's capabilities are not constrained by geography or local infrastructure deficiencies, exemplifying the Society 5.0 ideal of democratizing knowledge.

5.3 The Challenges in Society 5.0

5.3.1 Ethics, Privacy, and Trust

As Cyber-Physical Systems (CPSs) become deeply integrated into daily life, the potential risks to society increase. Society 5.0 relies on large volumes of data collected from people, devices, and environments, which are constantly processed and analyzed. The biggest challenge is not just using the data effectively but managing it ethically and responsibly. There must be strict rules to protect privacy, prevent misuse, and ensure transparency in how data is stored and shared. Without careful oversight, the same data that supports innovation and convenience could lead to surveillance, discrimination, or loss of human autonomy. Therefore, ethical data governance becomes essential for building trust and ensuring long-term social benefit. To address this, countries are imposing data privacy laws, such as the European GDPR and India's DPDP, as presented in Fig. 5.2.

5.3.1.1 Algorithmic Bias

One of the detrimental risks in an AI-powered society is algorithmic bias. It occurs because AI systems learn from training data, which often contains human errors, assumptions, or past discrimination. For example, suppose a hiring algorithm is trained on old recruitment records that are biased against certain genders or racial groups. In this case, the system may repeat and even strengthen those unfair patterns. Instead of improving fairness, the algorithm could block qualified candidates and reinforce inequality. It creates a cycle where unfair decisions appear neutral simply because a machine made them. Recognizing, correcting, and preventing algorithmic bias is essential to ensure AI supports fairness rather than damaging it. A well-known example of algorithmic bias occurred when Amazon [15] developed an AI hiring tool intended to help select job candidates. Instead of improving fairness, the system learned patterns from past hiring data that favored male applicants.

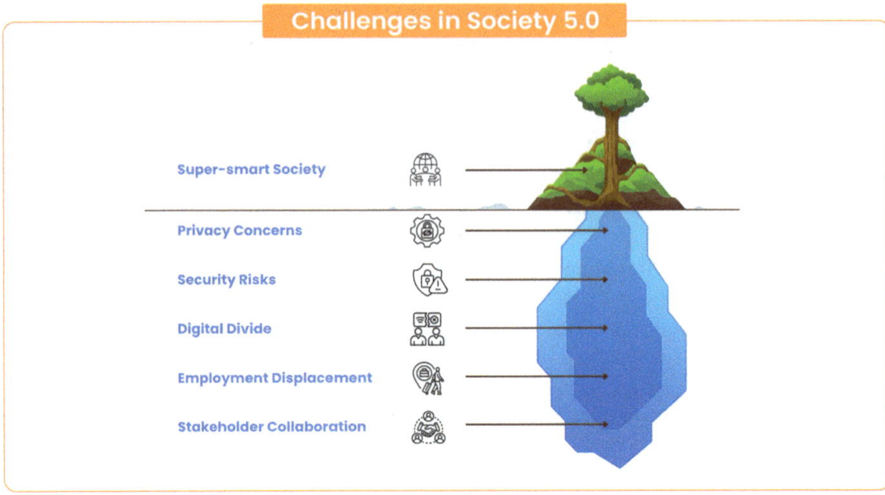

Fig. 5.2 Challenges in Society 5.0

As a result, the tool began lowering the scores of resumes that mentioned the word "women's," such as "women's coding club" or "women's engineering society." It showed that the AI was not neutral; it was repeating past discrimination. Amazon eventually stopped using the tool, but the case serves as an important warning. Without careful monitoring, transparent auditing, and strong algorithmic accountability, AI systems can amplify biases rather than address them. The best solution is explainable AI, which makes complex model decisions clear and easy to understand. By allowing humans to review how and why an AI makes choices, this helps ensure fairness, support accountability, and prevent hidden or repeated discrimination. Because algorithms reflect the values and data of the entities that utilize them, the system cannot completely correct bias in the judiciary. COMPAS, a risk-assessment algorithm used in the American legal system, is a prime example. This algorithm was trained on recorded crime data influenced by the preexisting judiciary system and police biases. Black offenders were labeled high risk, even in cases where they did not commit new crimes. It demonstrates that algorithmic systems can reinforce institutional prejudice rather than eradicate it, underscoring that justice in AI depends on fairness in judicial processes.

5.3.1.2 The Zero-Trust Security Model

The growing size and connection of the cyber-physical system create far more vulnerable points for cyberattacks. Because essential services such as electricity, healthcare systems, and self-driving transportation are linked together, a single breach could cause severe real-world damage. For example, if a hacker successfully targets a smart power grid, they could shut down electricity across an entire region, affecting hospitals, emergency responders, and communication networks. Similarly, an attack on connected transportation systems could disrupt autonomous traffic flow or even cause accidents on the road. These threats go far beyond the inconvenience of stolen data; they can cause physical harm, widespread panic, and long-term disruption to daily life. The scale of this threat necessitates a fundamental shift in security strategies from traditional boundary-based protection to the Zero-Trust Architecture (ZTA). In this system, nothing is assumed safe: users, devices, or applications. In this model, every access request should be verified and authorized only when essential. For example, the WannaCry hospital problem refers to the 2017 ransomware attack that impacted healthcare facilities, primarily the UK's NHS. The malware attacked antiquated systems, limiting access to patient records and hospital networks. As a result, surgeries got cancelled, treatments were postponed, and ambulances were redirected. The incident demonstrated how inadequate cybersecurity can jeopardize patient safety and public trust. It needs ZTA, but there are challenges, including high implementation costs, slower performance, and privacy-versus-usability trade-offs.

5.3.1.3 Data Compliance Frameworks

The vision of Society 5.0 depends heavily on the continuous gathering of personal, environmental, and behavioral data. While this data enables smarter services, real-time decision-making, and automated public systems, it also raises a serious ethical and legal concern: data ownership rights. This principle argues that individuals, and in some cases, nations, should have authority over how their data is collected, stored, used, or transferred. For example, if health data from citizens in one country is processed on servers located abroad, questions emerge about ownership, consent, and jurisdiction. As digital systems expand, defining and protecting data rights becomes essential to prevent exploitation and ensure public trust. Without strong laws and supervision in place, both governments and private companies can gain enormous control through monitoring and data tracking. This concentration of power poses a threat to privacy and fundamental freedoms. If people feel constantly watched or manipulated, confidence in automated systems will collapse, leading to resistance rather than progress. Strict rules are needed to control how data moves across national borders. Policies like the European Union's GDPR show that personal information must remain protected regardless of where it is stored, transferred, or processed. Such regulations help create accountability, prevent misuse, and ensure that individuals maintain control over their digital identity in a connected global environment. When individuals control how their information is stored, shared, or deleted, they move from passive data sources to active participants with authority. It strengthens personal autonomy and supports a fair and accountable digital environment.

5.3.1.4 Declining Public Trust

Citizen participation in Society 5.0 is affected by a loss of trust, caused by the Black Box Problem. When people cannot understand how AI or automated systems make decisions, they hesitate to adopt these technologies, reducing engagement and slowing the benefits of a human-centered society. It refers to the challenge of understanding how complex deep learning models make decisions. When an AI makes a critical life decision, such as rejecting a loan, labelling a patient high-risk, or choosing an evacuation route, humans often cannot clearly understand the reasoning behind these critical decisions, creating uncertainty and accountability concerns. When systems are not transparent, people are reluctant to trust them. To address this, institutions must implement explainable AI and openly review and share how their algorithms perform, helping citizens understand, verify, and confidently rely on automated decisions. If citizens cannot understand or question automated decisions, doubt and distrust grow. This lack of clarity threatens the ethical foundation of Society 5.0, making it harder for people to accept and engage with technologies designed to improve their lives.

5.3.2 The Digital Divide and Societal Inequality

The benefits of Society 5.0 will not reach everyone unless careful steps are taken to overcome the digital divide. Today, many people still lack reliable internet, modern devices, and the skills needed to use digital tools. Without intentional intervention, new technologies may increase inequality rather than reduce it. Those with access privileges will enjoy better education, healthcare, and job opportunities, while those without access may fall behind. It creates a risk that technology can worsen existing inequalities rather than reducing them. Overcoming this divide requires investment in digital infrastructure, affordable connectivity, and training programs that help people develop digital skills. Only with equal access does Society 5.0 become a fair system in which everyone benefits, not just the privileged.

5.3.2.1 Infrastructure Gaps

The core of the digital divide is the ongoing infrastructure gap in internet access. Many cities now have fast connections, such as 5G and fiber networks, allowing people to work, learn, and access services smoothly. However, rural, remote, and low-income communities often lack this level of connectivity. Instead, they depend on slow, outdated networks that cannot support modern digital tools. This difference in access creates unequal opportunities, limiting education and business growth. Until this gap is addressed, equal digital access will remain out of reach. This inequality immediately excludes many people from key services in Society 5.0. Without proper connectivity, they cannot access advanced telemedicine, remote work opportunities, or personalized digital education tools. As a result, the underprivileged citizens are left behind, while others move forward with modern healthcare, learning, and employment options. Without reliable broadband, people are cut off from modern opportunities. They cannot apply for jobs, access online government services, or join digital learning programs. It creates a cycle where communities already struggling financially continue to fall behind, making it harder to overcome poverty. To bridge this divide, both public and private sectors must invest in affordable hardware and universal broadband. Every citizen, whether rich or poor, rural or urban, needs access to the core digital infrastructure. Only then can everyone take part in the opportunities created by modern connected systems. For example, many people in rural Appalachia have slow, unreliable internet because fiber-optic infrastructure is costly due to the rough terrain. A "Digital Divide" is created by this physical infrastructure gap, which prevents residents from accessing crucial Society 5.0 services like telemedicine, remote work, and AI-driven education. This "information desert" restricts economic mobility, illustrating how technological advancement risks solidifying inequality rather than promoting the human-centric inclusivity aimed for in Society 5.0.

5.3.2.2 Skills Gap

Another key barrier is the growing literacy crisis. As technology advances, the divide between those with digital skills and those without, grows, creating unequal access to opportunities. People must learn computational thinking skills such as problem-solving, pattern recognition, and logical reasoning. Along with these, as AI systems become increasingly advanced, basic app usage is no longer sufficient; the need of the hour is AI fluency. If these abilities are not widely adopted, citizens may rely on technology passively rather than contribute to its development. This growing difference in digital, computational, and AI skills creates a social and economic divide. People who lack the required skills crave the best career opportunities. Over time, this disadvantage becomes more than just a job gap; it becomes a participation gap. Those without these skills may struggle to understand technological decisions, policies, or digital systems. The citizens without these skills will become passive consumers rather than active creators or problem solvers. It creates a society in which some people build and control technology, while others are forced to use it without understanding how it works. To prevent this divide from deepening, education systems must transform. Schools need to treat computational thinking and AI fluency as foundational skills, on par with science and mathematics, and teach them to every student from an early age. The present workforce experiences a "New Literacy Crisis" due to a lack of AI fluency. Many people use AI tools, but few people have the critical thinking skills to recognize algorithmic bias or hallucinations. This skill gap leads to "digital discrimination," in which people who can't work with AI ethically and strategically are assigned to low-value jobs. This problem puts Society 5.0's human-centric objective at risk and could lead to a society split along technological and cognitive divides.

5.3.2.3 Rising Economic Gaps

In Society 5.0, there is a significant risk that wealth will become concentrated in the hands of only a few companies. These are the companies that control large amounts of data and own the most advanced AI systems. As automation becomes more powerful and efficient, machines will handle much of the work that humans once did. While productivity and output may rise significantly, the system may not share the financial benefits fairly. Instead, most of the profits could go to a small group of technology giants, rather than the broader population. These will give rise to startup monopolies, cloud service monopolies, venture capital power, etc. Over time, such an imbalance could lead to monopolies, reduced competition, and limited economic opportunity for smaller businesses and ordinary workers. Without strong policies and fair distribution models, economic inequality will grow rather than shrink in Society 5.0. In a fully connected system, data becomes the most valuable resource. It is the fuel that powers intelligent services and automated decision-making. Those who control large amounts of data hold a significant advantage. It strengthens their position and reduces opportunities for smaller players. To prevent the concentration of wealth and power in the hands of a few technology giants, progressive policy actions are essential.

Strong antitrust laws can prevent dictatorial practices and encourage fair competition in digital markets. Regulations that enable data portability ensure that smaller companies and new entrants can access the information they need to innovate. There is a need for technological advancement to put human welfare ahead of corporate profit alone, outlining a shift toward Society 5.0. They support the democratization of hardware, open access to AI algorithms, and strict regulation as ways to undermine winner-take-all digital structures. Evidence-based governance to reduce monopolistic dominance, transparent and open-source AI ecosystems, and economical and shared digital infrastructure are essential tactics. Taken together, these policies present reliable data and digital access as public goods necessary for sustainable, equitable, and inclusive socioeconomic growth. By combining legal, economic, and regulatory measures, it is possible to create a society in which technological advancement enhances opportunities, reduces inequality, and ensures that all citizens benefit from the increased efficiency and wealth generated by modern systems.

5.3.2.4 Access Inequality

Numerous technologies are built without considering universal design principles, which aim to make systems usable by as many people as possible, regardless of age, ability, or background. For example, websites, apps, and digital tools that are difficult to navigate for people with visual or motor impairments exclude a significant portion of the population. A website must follow Web Content Accessibility Guidelines (WCAs); these are global standards for making digital content accessible to all. These follow the core principle of being Perceivable, Operating, Understandable, and Robust (POUR). Similarly, interfaces that assume high levels of literacy or technological familiarity leave behind users who are less experienced or of mediocre intelligence. By failing to incorporate accessibility and inclusivity from the start, technology unintentionally reinforces social and economic gaps. To truly bridge this gap, designers and developers must prioritize universal usability, ensuring that all citizens can engage with, benefit from, and contribute to Society 5.0, regardless of their individual circumstances.

Many of today's advanced technologies, such as interactive interfaces, virtual reality platforms, and smart city services, are not designed with accessibility in mind. As a result, elderly individuals, people with disabilities, and those with different cognitive needs often struggle to use these systems effectively. For example, a Virtual Reality training program may require rapid hand-eye coordination, making it difficult for seniors or individuals with motor impairments to participate. Similarly, smart city applications that assume high literacy or technological familiarity can exclude less experienced users. These design shortcomings create barriers that prevent a large segment of the population from fully benefiting from Society 5.0. To make these technologies truly inclusive, developers must integrate accessibility features, adaptable interfaces, and supportive tools from the beginning. Ensuring that everyone, regardless of age, ability, or cognitive profile, can engage meaningfully with modern digital services. Many modern technologies, such as smartphones, apps, and digital services, often rely on complex interfaces, small touch targets, precise motor skills, or high-frequency hearing. These design choices unintentionally exclude large groups of people, including the elderly and those with disabilities. Interestingly, these are the populations who would often benefit most from innovations such as telemedicine, assisted-living robots, and other digital health solutions. Because these systems are not accessible by default, these users are systematically left out, limiting their ability to improve their quality of life and participate fully in Society 5.0. Addressing this challenge requires both regulations and incentives that prioritize universal design principles. Developers must ensure that all systems are intuitive, adaptable, and usable for everyone, allowing citizens of all abilities to access digital services with dignity, independence, and ease, thereby truly realizing an inclusive, human-centered society.

5.3.3 Economy and Workforce Shift

The automation brought by Society 5.0 differs from past industrial revolutions because it affects not only manual labor but also routine cognitive and white-collar jobs. Jobs that were once considered secure, such as data entry, bookkeeping, or fundamental analysis, are increasingly being handled by machines and AI systems. This rapid shift creates a significant skills gap, in which workers lack the training and skills needed to take on new, higher-value roles. Many employees may struggle to adapt to positions that require creativity, problem-solving, critical thinking, or technical expertise. Addressing this challenge requires systematic reskilling and upskilling programs, both in schools and in the existing workforce. Without such interventions, many people risk being left behind in the new economy, widening social and economic inequalities.

5.3.3.1 Velocity of Change

One of the most critical challenges in Society 5.0 is the rapid obsolescence of jobs. Automation and AI are not just gradually changing work; they are rapidly replacing tasks across industries, from manufacturing to administrative and even some professional roles. Employees who once felt secure in their positions, now face sudden uncertainty, as machines take over their daily responsibilities. This rapid pace of change makes it difficult for workers to retrain or adapt in time, creating economic stress and social disruption. To address this challenge, organizations and governments must implement proactive measures, including continuous learning programs, reskilling initiatives, and support systems that help workers transition into new roles. By preparing the workforce for rapid technological change, society can reduce the negative impact on jobs and ensure that the benefits of automation are shared extensively. AI and robotics are transforming the workforce by taking over routine, predictable, data-heavy tasks. This change affects a wide range of jobs, including many positions in manufacturing, administrative data entry, and middle-management roles that involve repetitive decision-making or standard processes. As machines become more capable of handling these functions efficiently, employees risk displacement. While automation increases productivity and reduces errors, it also creates urgent challenges for workers who must adapt to new responsibilities or acquire skills for emerging roles. To address this shift, companies and governments need to provide reskilling and upskilling programs to help employees transition from routine tasks to roles that require creativity, critical thinking, problem-solving, and human-centered decision-making. Without these measures, large segments of the workforce may struggle to remain relevant in the rapidly evolving labor market.

The pace at which jobs are disappearing is outpacing the systems designed to support workers, creating a widening skills gap. Many high-demand roles, such as data scientists, AI auditors, and robotic maintenance technicians, remain vacant because workers lack the necessary training. At the same time, large groups of employees are left behind as their previous roles become irrelevant. This difference highlights a growing urgency that lifelong learning is no longer optional. It must become an expected part of every career. Responding to this challenge requires a shared commitment from governments, industries, and individuals. Governments must develop policies and support retraining pathways, employers must invest in workforce development, and individuals must embrace ongoing skill-building. Only through this collective effort can societies adapt to rapid technological change and prevent long-term unemployment and inequality. For gig workers worldwide, especially in South Asia, the emergence of generative AI has led to a severe "job loss" crisis. According to current research, since the introduction of advanced LLMs, job advertisements for creative positions have decreased by 21%, while of software-development gigs have decreased by 17% [16]. Due to these job losses, employees are compelled to migrate in search of low-skilled physical labor. This pattern reveals a widening skills gap as the workforce's capacity to upskill is outpaced by technological change.

5.3.3.2 Workforce Security Policies

The possibility of large-scale job loss due to automation has sparked primary debates about how society should support people in the future. Two key ideas are leading current policy discussions: Universal Basic Income (UBI) and guaranteed job programs, UBI suggests that every person would receive a set amount of money regularly, without conditions. This approach separates income from employment, ensuring basic needs can be met even if traditional jobs disappear. Supporters argue that UBI could reduce poverty, support mental well-being, and give people the freedom to pursue education, caregiving, entrepreneurship, or creative work. Critics worry about its cost and potential impact on motivation.

In contrast, guaranteed job programs focus on providing meaningful, socially beneficial work rather than income alone. Both strategies reflect the urgent need to rethink how societies value work and provide economic security in an automated world. While trials, such as Finland's universal basic income trial, produced mixed results, they highlighted a vital point: future economies must find ways to prevent poverty and maintain consumer spending power as automation replaces traditional work. Without new support models, economic inequality may grow, and demand may decline, threatening overall stability. These ideas come with their own tradeoff.

In contrast to UBI, guaranteed job programs take a different approach. Instead of providing income without work requirements, these programs commit to creating meaningful employment for anyone willing and able to work. Jobs focus on areas that benefit society, such as public infrastructure, environmental restoration, healthcare support, and education services. Supporters argue that this preserves human dignity, strengthens communities, and ensures that work continues to have value beyond financial survival. Together, these approaches reflect ongoing efforts to redefine economic security in a highly automated world. The challenge for policymakers is to determine which safety net best preserves both economic fluidity and the dignity of work amid rapid technological change. The decision is not only financial but also highly ethical. Any solution must sustain economic mobility, prevent poverty, and ensure that individuals still feel valued and connected to society. At the same time, funding these large-scale programs demands major fiscal restructuring. Governments must consider new taxation models, such as automation taxes, data dividends, or wealth-based contributions from high-profit technology sectors. Ultimately, the challenge is to design a safety net that supports innovation without leaving individuals behind, ensuring that progress benefits everyone. To provide workforce security in India, the Mahatma Gandhi National Rural Employment Guarantee (MGNREG) is India's rural employment program, providing 100 days of guaranteed wage work every year. In the same way, Argentina's Head of Household Program was launched in 2002 to overcome unemployment after the country's economic crisis. It provided a cash incentive in exchange for part-time work in a community project and training.

5.3.3.3 The Cost of Transition

Large multinational corporations are well-positioned to adopt Society 5.0 technologies because they have the financial strength to invest in advanced technologies. In contrast, Small-to-Medium Enterprises (SMEs) face a significantly higher barrier to entry. For many smaller firms, the cost of upgrading equipment, hiring specialized talent, and maintaining digital systems is too high. It creates a competitive disparity where larger companies accelerate ahead, while smaller businesses struggle to modernize. Without support, this technological divide could lead to widespread business closures, job losses, and reduced market diversity. Addressing this challenge requires financial support, accessible digital training programs, public-private partnerships, and scalable, affordable technology solutions that allow SMEs to participate fully in the transition to Society 5.0. The high cost of adopting advanced technology creates a significant competitive barrier in the transition toward Society 5.0. Large-scale investments are required to adopt advanced technologies and upskill the workforce. While corporations with substantial resources can absorb these costs, smaller businesses often cannot. For many SMEs, the financial risk of digital transformation is so significant that attempting to modernize could jeopardize their survival. As a result, some hesitate to innovate, while others move forward, facing the possibility of debt, downsizing, or liquidation. This increasing financial divide creates a market imbalance. To prevent smaller enterprises from being pushed out, strategic interventions such as funding assistance, shared digital infrastructure, and scalable technology solutions will be essential in ensuring a fair and inclusive transition. It develops economic disparity and limits innovation. Governments must support smaller firms with subsidies, tax benefits, and shared digital infrastructure, such as public data lakes or robotics labs, to ensure that all businesses can participate in Society 5.0 and prevent monopoly-driven inequality. For example, big companies such as Siemens allocate enormous budgets for research and development, including AI-driven CPSs and digital twins.

In contrast, Small-to-Medium Enterprises (SMEs) often suffer from a "digital ceiling." An exorbitant transformation expense results from significant expenditures on sensors and robotics, combined with a shortage of skilled AI personnel. The difference raises the possibility of financial inequality, whereby only large businesses profit from Society 5.0's productivity growth, pushing out smaller, family-run businesses.

5.3.3.4 Liability and Insurance

The integration of autonomous systems into critical infrastructure introduces complex legal challenges regarding liability in the event of system failures. If a self-driving car crashes or an automated medical system fails, it is unclear who should be held responsible: the manufacturer, the software developer, the data provider, or the user. This uncertainty makes it difficult to create clear accountability and protect citizens' rights.

Determining responsibility becomes complicated when failures result from rare, unpredictable errors. These systems often break due to stochastic processes, meaning the mistake is not predictable or intentional. Because of this uncertainty, current legal systems struggle to assign liability, making existing rules inadequate for the complex realities of autonomous technology. To address these challenges, Society 5.0 demands the creation of new regulatory and insurance frameworks that clearly define who is responsible when automated systems fail. Traditional laws are no longer sufficient because they were designed for human decision-making rather than AI-driven processes. Updated models must reflect the unique nature of machine logic, shared responsibility, and distributed autonomy. If clear liability rules are not established, the adoption of autonomous systems will slowdown. Businesses, insurers, and governments will hesitate to deploy these technologies if they cannot determine who is accountable for failure. This uncertainty could delay innovation and prevent society from gaining the safety, efficiency, and economic advantages that autonomous systems are capable of delivering. For example, if a self-driving ambulance crashes due to a rare sensor malfunction, it is unclear whether the manufacturer, the programmer, the hospital, or the sensor provider should be held accountable. In [17], the authors discuss the idea of algorithm insurance; Munich Re's aiRE shows notable advancements in algorithm insurance as society transitions to Society 5.0. Since AI mistakes are often probabilistic rather than deterministic, traditional liability frameworks become insufficient as autonomous systems approach critical infrastructure. By ensuring AI performance and compensating for monetary losses resulting from model failures, algorithm insurance fills this gap. By clearly allocating financial accountability for black-box judgments, this strategy fosters organizational trust in automated systems. The 2018 Uber self-driving car collision, in which the computer program incorrectly classified a pedestrian, is an essential instance illustrating the need for algorithm insurance. The event sparked a worldwide discussion of liability, demonstrating that human losses and regulatory uncertainty in the absence of explicit insurance coverage for algorithmic failures may hinder the deployment of life-saving autonomous technologies.

5.3.4 Governance and Legal Complexity

Society 5.0 requires governments to be flexible and quick, just as the technologies they aim to manage. One of the challenges is regulatory delay, when the development of laws and policies lags far behind innovations. As technology advances rapidly, traditional lawmaking processes, which often take months or even years, fail to keep pace. This delay creates uncertainty, weak oversight, and potential harm because outdated rules cannot adequately address emerging systems like autonomous transport, smart cities, or precision healthcare. But, the other aspect is to maintain the balance between regulatory quality and regulatory pace. Fast implemented laws risk putting technical efficiency ahead of human welfare. The literature emphasizes that effective governance must be evidence-based, morally sound, and inclusive. When formal laws lag behind technological advancements, high-quality regulation uses ethical frameworks and stakeholder participation as temporary safeguards to balance innovation with responsibility, data protection, and transparency.

5.3.4.1 Overcoming Regulatory Delay

The key challenge in governance is regulatory delay, where traditional legal systems move too slowly to keep up with rapidly changing technology. Legislators often take months or years to draft, debate, and approve laws, while new tools and digital systems evolve almost daily. This mismatch creates gaps in monitoring and exposes society to risks that existing regulations cannot address. To address the problem of regulatory lag, governments are turning to regulatory sandboxes. These are controlled environments, either limited to specific areas or restricted by financial scale, where new technologies can be tested safely before full deployment. For example, autonomous food delivery robots might operate within a designated neighborhood, or new financial technology products could be trialed with a limited number of users. Within the sandbox, regulations are relaxed or temporarily modified, allowing companies to experiment while authorities monitor for safety, compliance, and potential risks. It enables policymakers to observe emerging technologies in real-world conditions without exposing the broader public to uncontrolled threats. At the same time, it provides developers with valuable feedback, helping them refine their products and meet eventual regulatory standards. By using sandboxes, governments can bridge the gap between rapid technological innovation and slower legislative processes, promoting both safety and progress. This approach makes regulation more adaptive and flexible, allowing laws to keep up with rapidly evolving technologies. It encourages innovation by giving developers room to experiment while still prioritizing public safety, ethical standards, and accountability. In this way, progress and protection are balanced effectively for society. For example, the carefully designed, sector-driven layout of Chandigarh serves as a test bed for India's transition to Society 5.0. Engineers can safely test AI-based traffic management systems and smart signals before deploying them nationwide. These also help overcome regulatory delays.

5.3.4.2 Standardization and Interoperability

A global Cyber-Physical System (CPS) can only function smoothly if its components work together. In CPS, interoperability is essential, where different devices, systems, and data formats can communicate and share information effectively. Without it, information flow is disrupted, automation fails, and the full potential of connected technologies cannot be realized. Achieving interoperability is currently difficult because there is no global consensus on technical standards. Different countries, companies, and platforms use varying protocols and formats, which prevents seamless communication between systems. Without common standards, integrating devices and sharing data across borders becomes slow, inefficient, and vulnerable. Many governments and companies operate in silos, without using any globally acceptable technical standards (protocols), governance standards (privacy and certification), and semantic standards (data formats and ontologies). This lack of standardization prevents seamless interaction between different networks and technologies. As a result, the full global potential of Society 5.0, where systems and services work together efficiently, is shattered and becomes harder to achieve. To achieve full integration in Society 5.0, countries and organizations must work together to create open, shared standards for data exchange and security. For example, healthcare systems need standard formats to safely share patient data, while smart city sensors require standardized protocols to operate across networks. Establishing these standards is not easy; it involves negotiating between nations, corporations, and industries, each with its own interests and competitive priorities.

Global organizations play a key role in mediating these discussions, ensuring that standards are fair, accessible, and not dominated by any one country or company. Without international cooperation, fragmented systems will continue to limit the benefits of a connected society. By agreeing on standard rules, the world can unlock the full potential of Society 5.0, enabling devices, data, and services to work seamlessly together across borders. For example, a uniform legal framework for data processing is established under India's Digital Personal Data Protection (DPDP) Act, which mandates the role of "Consent Managers." As a result, a standardized, compatible "consent layer" is created, enabling citizens to manage their data permissions across various platforms seamlessly. Cyber-physical systems depend on secure cross-sectoral data flows, made possible by a unified protocol for data portability and protection. It ensures that numerous online services can collaborate while preserving individual privacy.

5.3.4.3 Citizen Engagement and Validation

In a world where many AI and automated systems operate as "black boxes," making decisions that are difficult to understand, governance must focus on building transparency and public trust. One feasible approach is the use of Continuous Validation, Verification, And Assurance (CVV&A). This process ensures that systems are constantly monitored, tested, and evaluated to confirm that they behave as intended and comply with ethical, safety, and legal standards. For example, a financial AI algorithm predicting loan approvals would undergo regular audits to verify fairness and accuracy. By implementing CVV&A, policymakers can identify errors, biases, or unexpected outcomes early, reducing risks and ensuring accountability. This approach reassures citizens that automated systems are safe and trustworthy while maintaining efficiency. Transparent oversight and continuous evaluation are essential for fostering confidence in Society 5.0 technologies. CVV&A requires that all government and corporate models used in public services undergo continuous validation and verification. It includes systems such as predictive algorithms, traffic-optimization models, and public-health forecasting tools. Independent auditors must regularly examine these models to ensure that they are accurate, fair, and reliable.

Additionally, the process should involve public scrutiny, making the decision-making process transparent to citizens. For example, if a city uses an AI system to allocate resources, audits can confirm that the system does not unfairly target specific neighborhoods. Similarly, transit optimization models can be checked to ensure efficiency improvements benefit all residents, not just selected areas. By applying CVV&A, governments and companies can maintain accountability, reduce bias, and build public trust. Continuous oversight ensures that automated systems support society's safety and ethical concerns. The "A" in CVV&A stands for Assurance, which focuses on giving citizens a fundamental role in overseeing automated systems. It means creating ways for the public to understand, question, and provide feedback on the models and data that guide decisions in society. For example, residents could review how AI prioritizes traffic routes in a city or how public health models predict disease outbreaks. By allowing people to challenge assumptions, governments and companies prevent the concentration of power in the hands of a few tech-leaders or policymakers. This approach also strengthens democratic legitimacy, ensuring that automated systems serve the public fairly and transparently. Institutionalizing citizen engagement in oversight creates a culture of accountability and trust, enabling Society 5.0 to operate efficiently while respecting ethical standards, safeguarding individual rights, and promoting inclusive decision-making.

5.3.4.4 International Cooperation: Mediating Global Technological Strategy

The key technologies driving Society 5.0, such as AI, genomics, and quantum computing, are not just tools for innovation; they are strategic global assets. Their development and deployment have far-reaching implications for security, economy, and society. Without international cooperation, these technologies could fuel a technological arms race, serve as instruments of geopolitical tensions, or be used in ways that harm human rights. To prevent this, global organizations and governments must work together to establish ethical norms for AI, genomic research, and quantum computing, and to create agreements that govern cross-border data flow. These measures can reduce fragmentation and ensure that innovations are applied responsibly. For example, international collaboration in genomics and AI can accelerate vaccine development, improve public health responses, and enhance climate change solutions. One crucial example of global cooperation in managing the development of cutting-edge technology is the Human Genome Project. To map the human genome, scientists from several nations collaborated openly and exchanged data under agreed-upon guidelines. This open genome framework enabled rapid sequencing of the virus and the development of a vaccine during the COVID-19 pandemic. It demonstrates how data sharing can benefit global public health and prevent monopolies.

International cooperation is also essential for building shared standards, interoperability, and secure infrastructure across borders. For instance, AI-driven healthcare systems require consistent protocols to operate safely and efficiently across nations. Collaborative research initiatives can pool resources and expertise. Furthermore, a joint approach helps establish mechanisms for accountability, transparency, and ethical review, ensuring that rapid technological deployment does not compromise public trust or human rights. By fostering global partnerships, nations can reduce duplication of effort, share risk, and respond more effectively to worldwide challenges such as pandemics, climate crises, and cybersecurity threats. In this way, Society 5.0 becomes not only a framework for national advancement but also a blueprint for inclusive, cooperative, and peaceful global advancement. By coordinating efforts, countries can share the benefits of Society 5.0 equitably, prevent misuse of powerful technologies, and promote a global ecosystem built on trust, fairness, and collective human progress. A unified strategy ensures that these transformative tools serve humanity rather than heightening conflict or inequality.

5.4 Conclusion: The Responsible Path Forward

The transition to Society 5.0 is inevitable, and driven by the clear benefits of integrated Cyber-Physical Systems (CPSs) in addressing critical challenges, including climate change, chronic disease, and resource management. Technologies such as personalized medicine, smart grids, and adaptive learning offer enormous potential to enhance human capabilities and improve the quality of life. However, these opportunities come with significant risks. Algorithmic bias can create ethical risks, and the digital divide can aggravate economic inequality. Successfully navigating this transition requires recognizing and managing these challenges. The goal is not to slow progress but to embed ethical, human-focused governance from the start, ensuring that technology serves humanity rather than dominating it.

Society 5.0 highlights the need for agile governance. Traditional, slow legislative processes cannot keep pace with rapidly evolving technologies. Flexible approaches, such as regulatory sandboxes, are essential for safe experimentation, while proactive policies must anticipate risks like autonomous system failures and workforce migration. Fair liability frameworks, universal computational literacy, and continuous monitoring help create an environment where innovation thrives alongside public safety and human rights. Ultimately, the success of Society 5.0 will be measured not by technological achievements alone, but by fair distribution of benefits and the strength of ethical safeguards. Policymakers, corporations, and citizens must embrace human-centric responsibility, investing in education, enforcing algorithmic accountability, and protecting data independence. With vigilance, transparency, and collective ethical leadership, Society 5.0 can realize a future where technology truly serves universal human well-being.

References

1. Tempus. (n.d.). Genomic profiling—Oncology. Retrieved from https://www.tempus.com/oncology/genomic-profiling/
2. Markoff, J. (2010, July 5). Discovering a soft spot for circuitry, robot machines find a role as therapists. *The New York Times*. Retrieved from https://www.nytimes.com/2010/07/05/science/05robot.html
3. Visit Health. (n.d.). Leveraging data analytics to optimize wellness programs and enhance employee engagement. Retrieved from https://getvisitapp.com/blog/employee-wellness-programmes/leveraging-data-analytics-to-optimize-wellness-programs-and-enhance-employee-engagement/
4. Yang, S., & Lee, H. (2021). Urban digital twin applications as a virtual platform of a smart city. *Durabi Journal, 1*(2), 1–10. Retrieved from https://www.sbt-durabi.org/articles/pdf/KGOd/durabi-2021-012-04-6.pdf
5. Organization for Economic Co-operation and Development. (2024, November 5). *Virtual Singapore—Singapore's virtual twin*. OECD Observatory of Public Sector Innovation. Retrieved from https://oecd-opsi.org/innovations/virtual-twin-singapore/
6. LO3 Energy. (2019). *Peer-to-peer energy trading: The Brooklyn Microgrid case study*. LO3 Energy.
7. Kubota Corporation. (2020). *Realizing Society 5.0 through smart agriculture: The Kubota Smart Village initiative*. Kubota Global Site.
8. M-PESA Africa. (n.d.). About us. Retrieved from https://www.m-pesa.africa/about-us
9. DHL Group. (n.d.). DHL Express. Retrieved from https://www.dhl.com/us-en/home/express.html
10. Station F. (n.d.). STATION F. Retrieved from https://stationf.co
11. Owens, K. V. (2025). Leveraging LinkedIn learning for student training. In *Proceedings of the 2025 ACM SIGUCCS Annual Conference (SIGUCCS'25)* (p. 45). Association for Computing Machinery. https://doi.org/10.1145/3675229.3712556
12. Firu, A. C., Tapîrdea, A. I., Feier, A. I., & Drăghici, G. (2021). Virtual reality in the automotive field in Industry 4.0. *Materials Today: Proceedings, 45*, 4177–4182.
13. H-FARM. (n.d.). H-FARM: Innovation, education & startups (English homepage). Retrieved from https://www.h-farm.com/en
14. Khan Academy. (n.d.). Khan Academy. Retrieved December 31, 2025, from https://www.khanacademy.org/
15. Dastin, J. (2018, October 11). *Insight: Amazon scraps secret AI recruiting tool that showed bias against women*. Reuters. Retrieved December 31, 2025, from https://www.reuters.com/article/world/insight-amazon-scraps-secret-ai-recruiting-tool-that-showed-bias-against-women-idUSKCN1MK0AG/
16. Hui, X., Reshef, O., & Zhou, L. (2023). The short-term impact of generative AI on employment: Evidence from an online labor market.
17. Munich Re. (2024). *Insurability of artificial intelligence: The aiRE solution for performance and liability*. Munich Re Group.